Time and Space

By the Editors of Time-Life Books

TIME-LIFE BOOKS, ALEXANDRIA, VIRGINIA

Earth's Sacred Gateways

The creation of the world, the nature of time, humankind's role in the universe—people have contemplated these cosmic puzzles since life began. To understand how and why the world came to be, to fulfill the uniquely human need to impose order on chaos, early peoples formed myths and legends that wove the mysteries of origin and existence into comprehensible patterns of time and space. Integral to many of the traditional tales were sacred places—landscapes so extraordinary in shape and beauty that they inspired awe and reverence.

Some sacred places, such as Spider Rock in northern Arizona *(right)*, were held to be portals to other ages and dimensions of time. Others, including Mount Kailas in Tibet and Lake Titicaca in South America, were venerated within their far-removed cultures as the dwelling places of gods.

In some cases, dissimilar peoples came to think of their holy places in remarkably similar ways—sometimes even describing them with the same metaphors. American Indian, Asian, and Greek cosmologies have all, for example, designated their particular holy sites as the "navel of the world," the unifying center at the heart of all creation. Yet despite the attempts at explanation, time and space remain enigmas, perhaps never to be fully understood and accepted only on faith. If answers are to be perpetually elusive, then maybe the ancients were wise to choose as holy sites such splendid landscapes as those on the following pages; they are powerful earthly reminders of the world's greatest mysteries.

The Aerie of the Spider Woman

High atop the soaring spire of Spider Rock, say the Navajo of the American Southwest, the earth goddess Spider Woman makes her home. Her perch is a slim monolith of red sandstone that towers 800 feet above the floor of Canyon de Chelly in northeastern Arizona. The canyon has been inhabited, at one time or another, by the ancient Basket Maker peoples — the predecessors of the Pueblo Indians—and by the Navajo, the Hopi, and various small communities of cliff dwellers.

According to the mythology of the Hopi Indians, Spider Woman's abode guards the portal through which humans emerged from the womb of mother earth at the beginning of time. As related in one Hopi tale, Sotuknang, the god of the universe, created Spider Woman to be his helper on earth. As Sotuknang's deputy, she was thought of by the Indians as a wise and grandmotherly figure. She provided a link between the human and divine worlds, sometimes transporting people between the two spheres, carrying their souls in a basket spun from her web.

Ancient Greece's Navel of the World

As recounted in the classical mythology of Greece, Zeus—the king of gods and men—released two eagles from opposite ends of the world in order to find the center of his domain. The great birds traversed the breadth of the earth and met in the fertile valley of Delphi on the southern slope of Mount Parnassus. This spot was marked by a stone called the omphalos, which is also the Greek word for navel.

A temple to Apollo was built at this site and became home to the revered Oracle of Delphi. According to tradition, a prophetess called the Pythia would sit over a chasm in the ground beneath the temple, inhaling vapors that escaped the earth. Slipping into a trance, she would utter tales of events yet to unfold. Her words were then turned into verse by the priests of Delphi and carried the weight of divine revelation. Kings and commoners alike were known to consult the oracle, hoping to transcend the normal constraints of time by peering into the future.

The Holy Mountain of Tibet

In the clear, austere air of western Tibet's Kailas range—one of the highest, most remote and desolate regions on earth— stands Mount Kailas, long a holy place for the peoples of Asia. Around this peak revolve the rich and complex cosmologies of four distinct religions, whose pilgrims have journeyed to its ice-clad slopes for more than a thousand years.

The eleventh-century Tibetan saint Milarepa wrote, "The prophecy of the Buddha says, most truly, that this snow mountain is the navel of the world, a place where the snow leopards dance." To the Hindus, the Jains, and the adherents of the ancient pre-Buddhist Bonpo sect, Mount Kailas is the center of the world and a dwelling place of gods, where the temporal and the eternal unite.

Both Hindus and Buddhists believe that the mountain is the earthly embodiment of mythical Mount Meru, or "world mountain." The foot of this fabled peak, they say, rests in hell, while its summit occupies the highest realms of heaven. Hindus also teach that the sacred Ganges River springs from its summit to pour over the head of all-powerful Shiva, the creator and destroyer of all things, the god who transcends space as well as time.

Mystical Lake High in the Andes

On a wind-swept plain in the Andean highlands, straddling the border between Bolivia and Peru, lies Lake Titicaca, mystical birthplace of the Inca civilization. To the lordly Incas, the 3,200-square-mile lake—second largest in South America—was a place of rebirth and a milestone in a cosmology that envisioned time as a progression of eons, rising and falling like imperial dynasties. According to one version of the Inca creation myth, the legend of the Children of the Sun, our present epoch followed on the heels of a period of darkness and chaos on earth.

Observing humans living out their lives in a primitive and barbarous state, the sun god wept and took pity on them. He sent to earth a son and a daughter—Manco Capac and Mama Ocllo—to teach the human race to worship him and to adopt a civilized way of life. These saviors would teach humanity the arts of cultivating plants and grains. The sun god set down his children on an island in Lake Titicaca; from there they journeyed forth to establish the Inca Empire.

Australia's Shrine to Dreamtime

Looming 1143 feet over a flat scrub plain deep in the Australian Outback, Ayres Rock is a hulking dome of feldspar-rich sandstone, which glows fiery red in the setting sun. To the Pitjantjatjara Aborigines who inhabit this land, the furrows etched on the eroded surface of the monolith are like geophysical runes that tell of a long-ago period called Dreamtime.

This was the sacred mythical time when the world was being formed. Out of a featureless desert arose giant semihumans who roamed the earth, creating trails, building fires, hunting water, and fighting one another—as later humans would do.

When Dreamtime ended, natural features such as rocks, hills, and streams appeared on the landscape to record the events of the earlier age. The south face of Ayres Rock memorializes a battle between two tribes of snake-people, totemic ancestors of the Pitjantjatjara; the blood of a vanquished enemy now shows in a water stain on the face of the rock.

Such legends contribute to a cyclic understanding of time that links Aborigines to their Dreamtime ancestors in an endlessly repeating pattern. For them, past, present, and future are inextricably bound.

CHAPTER 1

Mythical Time and the Cosmic Order

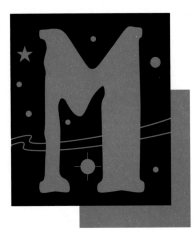

ichigan's *Dearborn Press* carried a report on May 10, 1973, of the following incident: While a woman named Laura Jean Daniels was walking home from work late one night, she looked up at the moon and reflected on the astronauts who had landed there. When she returned her gaze to ground level, her surroundings were no longer familiar. "Even the pavement on the sidewalk was gone," Daniels said. "I was walking on a brick path." The houses along the street had also disappeared, and several hundred feet ahead of her, at the end of the brick path, was a cottage she had never seen before. There was a heavy scent of roses and honeysuckle in the air.

Puzzled and frightened, she continued along the path. As she neared the cottage she saw a man and woman "in very old-fashioned clothes" sitting in the garden. "They were embracing, and as I drew closer I could see the expression on the girl's face," Daniels told the reporter. "Believe me, she was in love." Daniels, embarrassed to be witnessing so intimate a scene, was wondering how she could discreetly make her presence known, when a small dog came charging through a bush, barking madly.

The animal "was quivering all over," Daniels said. "The man looked up and called to the dog to stop barking, and asked him what he was barking at." At that point Daniels realized the man "couldn't see me . . . and yet I could smell the flowers and feel the gate beneath my hand." Confused, she turned and looked behind her—and saw her old familiar surroundings. "There was my street!" she related, "but I could still feel the gate in my hand." However, when she turned back to look toward the cottage again, it was not there. Instead, she said, "I was standing right in the middle of my own block, just a few doors from home." She never saw the cottage or the couple or the dog again. It was as if Laura Jean Daniels had briefly slipped through time, or space—or perhaps both.

Most of us rarely give a second thought to our everyday conceptions of time and space. Minutes and hours pass by—too slowly when we are waiting for a delayed flight, too quickly if we are hurrying to finish a task before a deadline. Day and night, summer and winter follow each other in unsurprising sequence. And the space we occupy has predictable and unyielding physical properties: When something is *here,* it cannot simultaneously be elsewhere. To get from one part of space to another—be it from this room to the room next-door, or from one planet to another—requires physical movement, or so we believe.

Once in a while, though, a Laura Jean Daniels will seem to step through time, or into another place or dimension, and—if we accept that the events happened as related—the framework we normally take for granted is suddenly knocked out of kilter. Skeptics usually dismiss such tales out of hand, saying that all of them involve fraud, hallucination, misunderstanding, or sheer ignorance, that none are truly unexplainable in straightforward physical terms. Other people are not so sure. They claim that close examination in many instances reveals no reasonable cause for the events described. They speculate about untested but conceivable explanations—that perhaps, for example, there are occasional fissures in time and space, faults through which people or objects might slip.

Certainly, stories about apparent displacement in time or space are particularly persistent in the literature of the paranormal. Over the years there have been many accounts of supposed time-slip experiences similar to that described by Laura Jean Daniels; they are generally categorized as retrocognitive events, from the Latin term for backward knowing. Of course, alleged acts of precognition—foreknowing—have been common features of human culture since the first seer told the first king what the future supposedly held. And there have been numerous reports, too,

of space, rather than time, being somehow transcended. These transcendences may take the form of sudden and mysterious vanishings of individuals or objects—sometimes, it is claimed, in plain view of witnesses—or of the corollary type of incident, when things or people seem to materialize where they are not expected to be.

Other examples of reputed time-space anomalies abound. So-called remote viewing, which would involve seeing across space rather than time, has even been the subject of military research by superpower governments, who would like to be able to spy on one another without leaving home. And many people believe rips in time or space, or in time *and* space, are the real explanation behind the scores of reported sightings of unidentified flying objects and supposed encounters with their alien occupants.

Ironically, it is the work of scientists—the very people who get most annoyed by what they regard as crank suggestions and paranormal prattle about quirks in space and time—that has fueled such speculation. In their twentieth-century quest for an elegant, all-inclusive understanding of the universe and its forces, astronomers, physicists, and mathematicians have proposed many theories that stand old beliefs on their heads and incidentally offer lots of room for kooks and serious paranormal researchers alike to play in without noticeably violating logic. Among these scientific theories: that space and time are not two different entities but aspects of a four-dimensional continuum that might better be called space-time; that this space-time can be bent or folded onto itself and that there may be holes in it; that more dimensions may exist than the four we are used

BAXTERS Patent Oil Printing 11 Northampton Square.

YGGDRASILL,

to thinking about, perhaps as many as twenty-six, although the extra dimensions may be curled up in tiny nodules we fail to notice; that instead of one universe there may be zillions of universes, all existing right here, right now.

Meanwhile, twentieth-century biologists, psychologists, and other researchers into human and animal structure and behavior have also been coming up with some startling notions relating to space and time: that many animals, perhaps including humans, have built-in compasses oriented to the earth's magnetic field or, in some cases, to the position of stars in the sky; that most living things have built-in biological clocks; that while some of those clocks are attuned to changes in daylight and darkness, others keep time according to mysterious internal signals and yet others seem to respond to the movement of heavenly bodies that the animals possessing the clocks cannot even see; that although all human beings are equipped with similar biological clocks, our perceptions of time and space depend greatly on our cultural environments and thus vary radically from society to society and from individual to individual. Some philosophical thinkers contend, in fact, that time and space exist solely as matters of human perspective.

All this thinking about the qualities of time and space is new only in terms of the specific ideas that emerge from it. People have been puzzling over the nature of those two elemental concepts since the human brain developed sufficiently to consider them. Many of the questions have remained the same through the ages: Do time and space exist outside our perception of them? Did time and the universe have a beginning, and will there be an end? Is the future already determined, or do our actions affect the way things will be? Where we look for the answers has changed, however. What became the stuff of philosophical and then scientific inquiry was at the outset solely a matter of religion.

Many of the eternal questions about space and time are encompassed in the subject of cosmology, the study of the evolution, structure, and laws of the universe. Nowadays, cosmology is fit work for astronomers and other scientists. In an earlier time, it was a subject for metaphysical philosophers. But originally it was the territory of shamans, priests, and storytellers, for the roots of cosmology are buried deep in the myths of ancient cultures. These intricate, often beautiful legends grew out of attempts to explain how and why the universe operated as it was observed to do, and just where humankind fitted into the plan. Each ancient culture described the world's beginnings with its own cosmogony, or account of creation.

In some cases, the explanations evolved by widely separated cultures bear striking similarities to one another. "Before Heaven and Earth had taken form, all was vague and amorphous," reads an anonymous Chinese account written down some 2,200 years ago. "In the beginning God created the heaven and the earth. And the earth was without form, and void," says the Hebrew Book of Genesis. "Where there was neither heaven nor earth sounded the first word of God," begins an old myth of the Central American Maya. "And . . . all the vastness of eternity shuddered."

lthough most creation accounts involved a divine motive force behind space and time, some of the schemes and mechanisms the old tales suggest provide startlingly close analogies to the theories offered by modern science. Anyone who believes, for example, that the universe is caught up in a series of big bang expansions alternating with cataclysmic contractions may find it eerily relevant that the Hindu god Shiva was thought to dance eternally, rhythmically creating and destroying and re-creating the universe to the beat of his drum.

For several early civilizations, the egg provided an ideal symbol for the birth of time and space. Here, too, in the effects of the eggshell splitting asunder, it is possible to find analogies to the big bang. According to Old Finnish lore, the teal—a kind of duck—laid its eggs on the knee of the Water Mother, the creation goddess. The eggs rolled off and broke, and their fragments became the earth, sun, moon, sky, and clouds. In ancient Greece, legend said that Nyx, the goddess of night, mated with the wind and hatched a silver egg,

Fortuna, the Roman goddess of good and bad luck, spins the Wheel of Time—also called the Wheel of Fortune—while seated on a red throne at the wheel's hub in this fifteenth-century illustration. According to legend, the wheel randomly carried some people up to success—such as the king seated at the top—and transported others down to misery, such as the wretch lying on the ground.

from which emerged Eros, the god of love. Eros pulled the rest of his shell away to reveal the earth and the sky. Traditional Samoans speak of their god, which they call the Heavenly One, as beginning his existence inside an egg. One day he pecked his way out; each shell fragment became one of the Samoan Islands.

Perhaps in order to elevate the importance of mere mortals in the larger scheme of things, the originators of the old myths often described cosmic events in terms of human behavior. Some ancient cosmogonies, for instance, were modeled on the process of sexual procreation. In one Egyptian myth, Atum, the creator, alone in the void, was forced to mate with himself in order to create a pair of fertile offspring, Shu (air) and Tefenet (water). That brother and sister then mated to beget Nut, the sky goddess, and Geb, the earth god, who were born embracing each other. Shu had

to force them apart to create heaven and earth. The Zuni Indians of the American Southwest reversed the Egyptian genders, so that the sky was male and the earth female. After coupling to create man and beast, the two separated, the earth sinking into the water, and the sky rising above the clouds.

Other creation myths seemed intended to account for nature's darker aspects. Ancient peoples were familiar with some animals killing others in order to eat, as well as with humans killing one another to gain territory, authority, or the favor of gods. Thus it is not surprising that many early cosmogonies incorporated symbols of death and dismemberment as necessary precursors to life. An old Scandinavian myth has it that the universe was built from the corpse of Ymir, a giant god of ice, who was murdered by his three great-grandsons. Ymir's skull became heaven; his flesh, the earth; his sweat and blood, the ocean; his teeth, boulders; his hair, trees; and his brains, the clouds.

A Sumerian myth relates an equally lurid genesis. Marduk, the son of gods, rose up against the ancient divinities, including his grandmother, the Mother Dragon, whom he stabbed in the heart. According to Sumerian scriptures, Marduk "split her into two like a dried fish. One half of her he set up and stretched out as the heavens," and the other half "he stretched out and made it firm as the earth."

Japan and Korea share a myth in which the abandoned wife of Izana-gi, the creator of

Eyes bulging and jaws agape, a giant crams a mutilated body into his mouth in Spanish artist Francisco de Goya's 1824 painting, Saturn Devouring His Children. Saturn was the Roman name for Cronus, the deity associated in late Greek mythology with time. He swallowed his children after being told that one of them would dethrone him. Only Zeus, who did indeed usurp his father's power, escaped the grisly fate. The painting symbolizes how time ravages and destroys all humans, who, unlike Zeus, cannot evade its clutches.

humankind, tried to get back at her ex-husband by eating 1,000 of her husband's offspring a day. Izana-gi thwarted her by creating 1,500 people a day, but he exhausted himself in the process and drowned in a stream, only to be reborn as the creator of the world. Note that in this case, humans existed before the world did, which is rare in mythological cosmogony. When the resurrected Izana-gi blew his nose, thunder and lightning broke out; his left eye became the sun and his right eye the moon. Grisly beliefs sometimes went hand in hand with grisly religious practices; if a deity had to die to make human life possible, then human sacrifices were necessary to honor that divine sacrifice.

Modern science, of course, admits of no morality in nature, but early peoples sometimes saw the whole of creation in terms of good and evil. The cosmogony embraced by followers of the ancient Persian faith Zoroastrianism was based on that idea. The religion's founder, Zoroaster, who probably lived in the seventh or sixth century BC, proclaimed that time was a cosmic battle between Ahura Mazda, the good Wise Lord, or Father of Greatness, and Angra Mainyu, the Prince of Darkness, who crept into living things and tried to destroy them. The Manicheans in the third century AD built their creation myth on the Zoroastrian tradition, but in reverse. The Manichean world was created out of the stuff of evil, but within each bit of matter was a pinpoint of divine light, waiting to be reunited with the divine. Manicheans tried to help this process along by abusing their own "evil" flesh.

No place on earth produced richer, more complex mythologies than the subcontinent of India, which gave birth to many versions of cosmic creation. The one that encapsulates the essence of India's primary religion, Hinduism, is found in the Veda, four sacred books composed in Sanskrit possibly as long ago as 1500 BC. Like Greek and Finnish creation myths, the Vedic story began with a golden egg, this one floating on an endless sea. After 1,000 years, the egg cracked open and the Lord of the Universe emerged to transform himself into the first man. This first man was lonely and afraid, so he split in two to create man and woman. From this original couple came every animal, fish, bird, and insect on earth.

The meaning of this myth reveals an attitude toward space—that is, toward the whole of the physical universe—that stands in direct opposition to what most Westerners believe. Yet it is the fundamental message of Hinduism: Although the world appears to teem with a multitude of beings and objects, they are all merely illusory manifestations of one god. To some Hindus, this universal deity is Brahma, who also appears as his alter egos, Krishna and Vishnu. Other Hindus regard either Krishna or Vishnu as the single, ultimate reality that lies behind everything else and view the two remaining divine incarnations as subsidiary. But whether they revere Brahma, Vishnu, or Krishna as the universal creator, devout Hindus believe that nothing we normally take for granted as

As the Roman god of doorways, Janus looked both inside and outside and was therefore commonly portrayed with two heads, as on this Roman coin minted around 200 BC. Janus also presided over beginnings and endings, and at his main temple in Rome, the god's statue witnessed the beginning and ending of each day through two doors that opened on the eastern and western sides of the building.

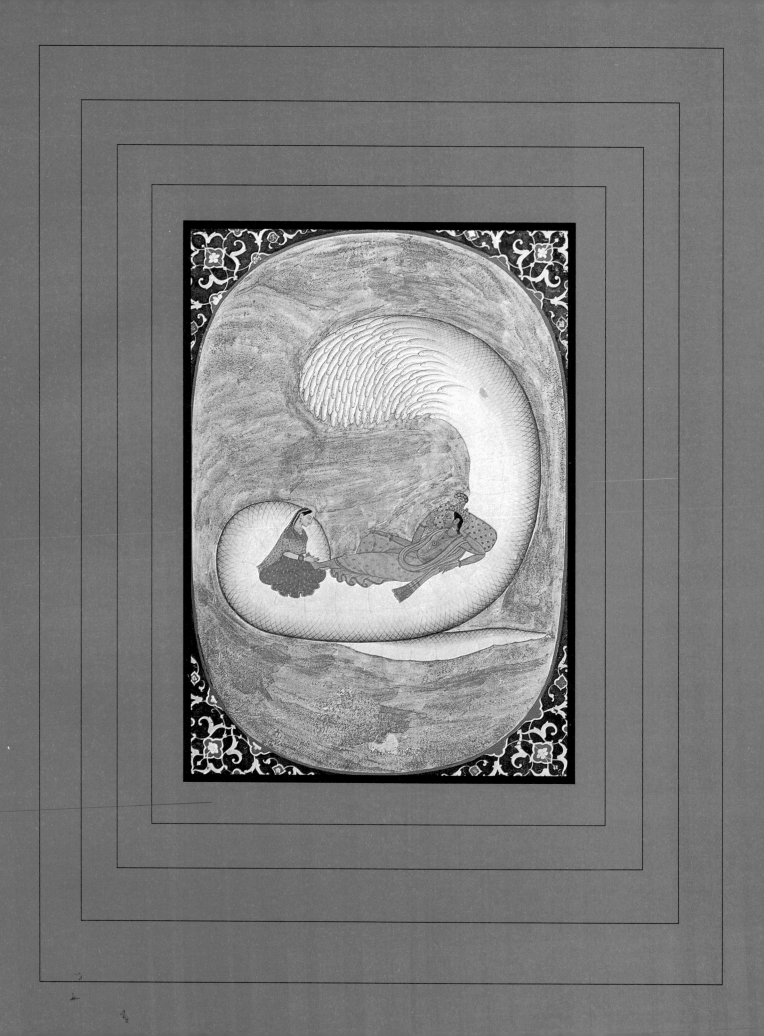

"real" exists. Anything that can be tasted, touched, seen, smelled, or heard is an illusion.

Moreover, Hindus consider time to be an illusion, a creation of the human mind, which is bent on dividing, categorizing, and misrepresenting time's real nature. Hindu mythology went to graphic extremes to bring this esoteric concept home. "Know that I am Time that makes the worlds perish when ripe and come to bring them to destruction," proclaimed Vishnu in the Hindu poem the Bhagavad-Gita. The ultimate spiritual goal of Hinduism is enlightenment, a high point of understanding from which the truth can be seen once and for all. Enlightenment means, among other things, escape from the illusion of time.

In the Hindu scheme of things, time and space—however illusory they may be—are cyclic in form. Shiva was often depicted in Hindu mythology as the Cosmic Dancer, whose rhythms followed the eternal cycles of space and time through repeated beginnings and endings, births, deaths, and rebirths. People proceeding through the cycles were reincarnated again and again, carrying their karma—the balance of good and bad deeds—from life to life, until they achieved spiritual perfection and graduated into timeless enlightenment.

In the meantime, according to Hindu myth, the universe is revolving through its own slow cycles. The time scale involved is truly cosmic. A year of a god is 360 human years, and 12,000 god years equals one blink of the god Vishnu's eye; a thousand of Vishnu's blinks—or 4.3 billion

Floating within the cosmic golden egg, the Hindu god Vishnu rests against Ananta, the many-headed serpent of infinite time, before initiating the creation of a new universe in this eighteenth-century Indian gouache. According to myth, the universe is constantly created and destroyed; before each successive creation, Vishnu's navel sprouts a lotus flower, from which the god himself emerges in his incarnation as Brahma, source of new life. Brahma breaks open the egg, thus beginning another cycle of rebirth and destruction.

Krishna, one of the incarnations of the god Vishnu and a symbol for the infinite space of the cosmos, reveals his cosmic form to his disciple Arjuna in this eighteenth-century Indian gouache. "Behold now," Krishna says to Arjuna in the classic Hindu poem the Bhagavad-Gita, "the entire universe with everything moving and not moving here, standing together in my body."

human years—is a single cycle of the world's creation and destruction. In fact, that figure is not far off modern science's best estimate of the world's age—four billion years or so. It seems almost uncanny that some 3,500 years ago the Aryan cattle herders who created the Hindu myths could have conceived of so great a span of time as 4.3 billion years, much less arrived at a figure that bears even a coincidental similarity to the age of the earth.

Later, possibly discouraged by the vast scale of time involved in working their way toward enlightenment, some Indians developed alternative religious ideas that allowed believers to get off the world's wheel sooner rather than later. The Jainist sect, founded in the sixth century BC, developed some of these new notions.

The Jains had no creation myth because its adherents believed there was no such thing as a beginning to time or space. Instead, the Jains saw time as a wheel with twelve spokes. Each spoke represented an age of the world. At the top spoke of the wheel, life hit its apex in an age called Very Beautiful, Very Beautiful. Humans grew to be six miles tall and had 256 ribs. All people were born as twins, who married each other and lived for thousands and thousands of years. The world was made of sugar, wine ran like water, and trees bore everything from fruit to precious jewels. That age lasted for no fewer than "400 trillion oceans of years."

Unfortunately, the quality of life diminished as the wheel revolved downward. The next age was only Very

Beautiful, and life was just half as grand. People grew to be merely four miles tall and had but half as many luxury-bearing trees to meet their needs. The ages descended from bad to worse; in the lowest age people were a scrawny eighteen inches in height, had only eight ribs, and lived to be no older than twenty. The days were scorching and the nights freezing.

The good news about this final age was that it lasted just a few thousand years. Then, after seven days and nights of rain, dwarflike humans emerged from underground to begin an ascent that took "ten millions of ten millions of one hundred millions of one hundred million periods of countless years," to reach the top once more.

This Jainist time frame was accompanied by a carefully defined image of space: a universe in the shape of a standing female colossus. Earth lay along a plane intersecting her waist, and seven continents and seven seas grew outward from her center in concentric rings. One of the seven land-masses was called Continent of the Purple Willow, and one of the seas was called Ocean of Clarified Butter. Seven levels of hell intersected her pelvis, legs, and feet; fourteen levels of heaven rose in planes through her chest, shoulders, neck, and finally through her head, which was surmounted by a golden umbrella of heaven.

Through this cosmic matron wandered the souls of humans, animals, and objects in various stages of enlightenment. As with Hinduism's karma, bad deeds weighed souls down and good deeds drew them up. Their spiritual progression paralleled their physical evolution from dust, to salt, to weather, to plants and trees, to fire, to insects, to fish, to mammals, to birds, to man, and eventually to the divine ranks of the Fully Realized, at which point they finally escaped the endless round of reincarnation.

But the lure of Jainism was that adherents could hurry this process along by rejecting as much of earthly life as possible. They embraced excruciating asceticism. Jainist monks and nuns begged for their food, pulled their hair out with their bare hands, and in many cases starved themselves to death. The cult's disdain for life on earth was embodied by

the Jains' savior, Lord Parshva. The archetypal ascetic, Parshva went without clothes and consumed very little food for twelve years. He grew so accustomed to discomfort that his wracked body remained absolutely motionless, even when a demon sent elephants, scorpions, tigers, a cyclone, and a fire-breathing monster to disrupt his meditations.

To cope with the cruel cycles of death and rebirth, the Jains went to extremes. Other cultures found ways to live more comfortably with the idea of cyclic time.

Ancient China, separated from India by both distance and geographical obstacles, was an inward-looking kingdom that took its own measure of things. Yet Chinese cosmogonies shared certain characteristics with their Indian counterparts. Again, it all began in an egg. In the Chinese creation story, the first being, called Pan Ku, was trapped in an egg surrounded by dark chaos. When he broke out, the top half of the egg floated up to become the sky, called yang, while the bottom half settled down to be the earth, called yin. For 18,000 years, Pan Ku held the two halves of the universe apart until they stayed that way on their own.

By then Pan Ku had exhausted himself. He lay down to die, and his body formed the known world: His limbs became mountain ranges on four sides (ensuring that the sky would stay up forever); the stars and planets were born out of his hair; one eye became the sun and the other the moon; his breath became clouds; and the parasites on his skin devolved into animal life. Nugua, the Chinese mother goddess, added human beings to Pan Ku's creation, forming them out of clay and breathing yin into some to make them females and yang into the rest to make them males.

The Chinese used their story of creation to determine the proper way to live, but unlike the Indians, their object was to enjoy life, not to escape it. Their creation myth introduced life's two quintessences: yin and yang. Yin represented "female" associations—moon, night, coolness, earth, passivity—while yang represented sun, day, warmth, sky, action. Yin was also identified with space, whereas yang was identified with time. The Chinese view of cosmic cycles was defined by the endless, rhythmic interplay between

A zerolike circle drawn by a Zen Buddhist symbolizes both the totality of the universe and the ultimate void, a dynamic, timeless emptiness from which all things come and whose nature is revealed through enlightenment. The characters at far left refer to a tea ceremony spiritually related to the symbol.

these two counterbalancing forces referred to as the Tao, the secret law that governed the cosmos. Unlike Hinduism, in which changes masked a still reality, Taoist philosophy asserted that all reality consisted of change.

The ultimate statements on the Tao were made by Laozi (Lao-tzu) in the sixth century BC. Laozi lived during a calamitous period of Chinese history in which political and military upheavals reduced society to near anarchy. Laozi and his followers turned away from the chaos to live as hermits, seeking an eternal order through individual contemplation. Nevertheless, Taoism was eventually adopted by Chinese society at large and has remained a hallmark of Chinese culture.

Whereas Westerners traditionally have tended to think of time, space, and objects *in* space as separate, unchangeable entities, Taoists saw all three joined inextricably in an ever-changing matrix (which, as it turns out, bears a somewhat closer resemblance to the modern scientific view of time and space as a composite whose quantities can change relative to each other). In Chinese art, clouds symbolize this perspective because they are constantly changing shape—quickly in a

This eighteenth-century Indian gouache depicts an ultimate transcendant state known to some Hindu and Buddhist followers as bliss-consciousness. Characterized by a sense of nonbeing, of existing outside space and time, and of being free from all thoughts, bliss-consciousness is supposedly achieved through the renunciation of the natural world, mortification of the physical body, and intense meditation. Disciples focus on the divine point of creation, which is thought to be the center of all experience as well as all being.

Zero: The History of Nothing

The concept of zero—the symbol that stands for nothing, or empty space—is often viewed simply in terms of its mathematical importance. Scholars debate who first brought zero into use; some say the Hindus, others the Babylonians, Chinese, Greeks, or Maya. But only the Hindus appear to have treated zero as a full-fledged number, dating back to AD 800. One reason the Hindus were comfortable with zero was that the symbol had a metaphysical and a mathematical connotation for them: Just as enlightenment was viewed as an empty space, yet dynamic and full of possibilities, zero represented nothing but could create other numbers.

Indeed, zero's connotations of nothingness, nonbeing, and infinity offended the rational minds of the Greeks, who did not find the concept logical. Conversely, the Hindus, who viewed nonbeing in a positive sense, as a step toward Nirvana, had no such qualms and fully pursued the mathematical possibilities of zero, which they called *sunya,* meaning "empty."

In the twelfth century, Arab scholars carried the idea of zero to western Europeans, who were using the laborious Roman-numeral system. Europeans at first found zero, with its implications of the void, unsettling, often regarding it as a device of the devil. But merchants adopted the new system because of ease in calculating, and soon the notion of zero found general acceptance, thus paving the way for the scientific and mathematical advances of the Renaissance.

The Directions of the Universe

As early humans sought to find order in the world, they had to learn concepts for orienting themselves in the space they inhabited. In addition to personal directions—in front of me, to her right or left—there were the unvarying orientations determined by nature. These depended on what part of the world the observer lived in, but in the Northern Hemisphere, they included east, where the sun rises, and west, where it sets; south, where the sun climbs highest, and north, where it never appears. Cultures the world over applied the concept of four directions to the universe in an effort to comprehend the idea of cosmic space. And just as an individual is the center of his own experience, they reasoned, so the universe must have a divine center from which all creation emanated—the locus of the directions.

The divine center and the directions that led to it held sacred importance for many cultures. For the ancient Aztecs, the cardinal points divide the world into four regions that represent the sequence of life itself, in which humans are born, live, weaken, and die. The locus symbolizes a fifth cardinal point; it is attributed to the fire god Xiuhtecutli, who symbolizes the hearth at the center of the house. And in Persian and Buddhist views of the cosmos, guardians of the four directions defend the world against demons.

Many cultures associated the four directions with sacred colors, which appeared in symbolic images reflecting the structure and stability of the universe. Some images, such as the mandala at lower right on the opposite page, were used as a focus of meditation by those who sought to stabilize their spirit. And Navajo sand paintings like the one at right were employed in ceremonial reenactments of the world's creation. A common goal of all the sacred images, however, was to orient humankind and the world in space.

This twelfth-century German illustration depicts man as a microcosm of the universe. Radiating from him are symbols of the elements—fire, water, earth, and air (clockwise from top)—from which the world was thought to have been created and which were said to correspond to the spatial directions.

A page from a pre-Columbian sacred book depicts the Aztec concept of the five cosmic regions (right). Anchoring the universe in the central region is the fire god; each region of the directions is represented by a tree symbolizing one aspect of the human journey—birth, life, death, and regeneration.

This Navajo sand painting depicts the Emergence Ladder, up which the first people climbed into the world. Colored bars symbolize the directions and figures in those colors signify the movement of the sun. Dawn Man stands on white, east; Twilight Man on yellow, west; Evening Girl on blue, south; Darkness Girl on black, north.

Flanked by saints and demons, a mandala symbolizes the structure of the universe in the fifteenth-century Tibetan Buddhist painting below. Four gates corresponding to the cardinal points lead from the heart of this meditation object to the rings, representing steps toward spiritual enlightenment.

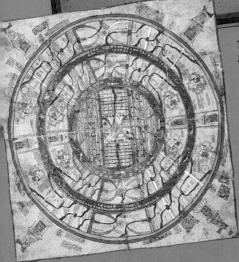

This seventeenth-century diagram from western India depicts the cosmos as expanding in four directions from the center of all, the mythical Mount Meru. Continents, countries, rivers, planets, and constellations are symbolically represented in the map; the four corners are marked by the deities of the four directions.

heavy wind, slowly on a calm day—but always changing. Taoists believed that even mountains were nothing more substantial than cloud forms, changing shape by the millennium rather than by the minute. They also believed that these outer changes corresponded to inner, personal ones. According to the Tao, you are changing even as you read these words; the information they convey becomes a new part of you.

Since only change was eternal, the best way for a person to live well was literally to "go with the flow." By heeding the patterns of change, one could follow them like a canoeist negotiating river rapids, a skill honed by balancing one's personal yin and yang. Unlike Hindus, who sought to escape human experience, the Chinese were immersed in life and worldly pleasures. (Chinese legends of those who had achieved Tao wisdom described them as living the high life in huge, richly furnished pavilions.)

Consequently, Chinese prescriptions for balancing spiritual energy could be quite literal. Elaborate Chinese sexual practices evolved in the belief that intercourse allowed partners to absorb each other's yin or yang in order to balance out their own. Some people tried to achieve internal balance by imbibing potions composed of substances with yin or yang character. A purple crystal called cinnabar was considered an especially potent harmonizer of yin and yang. Unfortunately, cinnabar also contained high concentrations of mercury, and a goodly number of Taoists were said to have died from poisonous overdoses of it.

The most profound approach to the Tao was through the *I Ching,* the ancient *Book of Changes* that predated Laozi and is the essence of the old, traditional Chinese culture and philosophy. The mythological "geometry" of the *I Ching* is deceptively simple: A solid line represents yang, and a broken line represents yin. To discern the present disposition and direction of the flowing current of the universe, the *I Ching* reader selected stalks cut from yarrow plants, or threw coins with lined faces, to create a random combination of six broken and unbroken lines, called a hexagram. There were sixty-four possible hexagram combinations, which together represented all the possible configurations of inner human and outer universal reality. Although latter-day enthusiasts frequently refer to the *I Ching* as an oracle, its original purpose was not fortunetelling. Instead, the reader was meant to gain a clearer perspective on his or her situation in the flow of time and space as it was occurring at that moment, in order to swim with the current of change.

Displayed against a background of hexagrams from the ancient Chinese text the I Ching, this badge from a nineteenth-century emperor's surcoat represents the nature of the cosmos and the balancing forces of yin and yang. The dragon symbolizes the dynamic, spiritual, time-oriented yang; the clouds and water surrounding it denote the passive, earthy, space-oriented yin. The pearl at the dragon's center represents the condensed energies of both yin and yang, which together are thought to generate the ever-changing universe.

According to one modern scholar of Chinese culture, Richard Wilhelm, "nearly all that is greatest and most significant in the three thousand years of Chinese cultural history has either taken its inspiration from, or has exerted an influence on the interpretation of its text." Many works of Chinese philosophy have addressed the meanings of each of the hexagrams, and the standard text that now accompanies the figures condenses the wisdom of the Chinese ages on the nature of "change in time."

If the ancient Chinese were reassured by changing time, the ancient Maya were obsessed by it. At the height of Mayan civilization, from the fourth to the eighth centuries AD, this Central American people ruled a domain that stretched across what is now Honduras, Guatemala, and part of Mexico. A rich corn economy enabled Mayan society to afford highly advanced, large-scale engineering projects, including an extensive highway system and the massive pyramids that are still standing today. In spite of these impressive accomplishments, the Maya—like the Nahua, another Central American people—devoted most of their resources and energies to tracking the ebb and flow of time.

The Mayan myths of creation are contained in the *Book of the Tiger Priests,* a volume in the *Chilam Balam of Chumayel,* a collection of sacred Mayan works. Here can be found the legend that begins, "Where there was neither

heaven nor earth sounded the first word of God." The story continues: "And He loosed Himself from His stone, and declared His divinity. And all the vastness of eternity shuddered." Despite the familiar Old Testament ring, this genesis has a telling difference: The Mayan creator is located *where* heaven and earth did not exist, not *before* they existed. This is because the Mayan people believed that time and space were one and the same, a single continuum.

Since time was not a separate entity to the Maya, they believed it could not move of its own accord but had to be conveyed by divine "carriers" who hauled the heavenly bodies through space. Like cosmic draft horses, these gods made time flow by pulling the sun, moon, stars, and planets through their celestial revolutions. "Good" time carriers brought good times, and "evil" carriers brought bad times. The main function of Mayan high priests was to predict what kind of times were coming by tracking the carriers' progressions, warning if an evil god was scheduled to "carry the day." They did so through highly sophisticated astronomical calibrations.

The Mayan obsession with the moving mechanism of time and space led them to produce the most elaborate and far-reaching calendar developed in pre-Columbian America. In addition to the 365-day year based on astronomical phenomena, the Mayan calendar recognized a sacred, non-

astronomical year of 260 days called a tzolkin, which was divided into thirteen 20-day units. Each of the 20 days in a unit was assigned to a god, as was each of the units themselves. By measuring a day god's attributes against the current unit's attributes, the priests determined the outcome of a particular day's events and activities. If, for example, the fourth day of a unit—always controlled by Kan, the corn god—fell in the tenth of the thirteen units, which belonged to Ah Puch, the god of death, a wise Mayan farmer would avoid planting his field.

The Maya constantly venerated the time carriers, since without proper encouragement, these hardworking gods might lay down their burdens and bring time and the universe to a dead stop. The Maya believed that the cosmic clock ran in fifty-two-year cycles, at the end of which the possibility of such a catastrophe was highest. It is said that the Spaniards waited until 1698 to invade Mayan territory because Spanish missionaries in previous expeditions had learned

A t'ai chi master in California (right) moves through the slow and precise patterns of that ancient Chinese art. T'ai chi is thought to balance the dancer's ch'i, or qi (vital energy), so that it harmonizes with the flow of qi in nature, thus aligning the practitioner with the natural rhythms of time and space. It is believed that buildings, too, need to be aligned with nature's energy. The purpose of the geomancer's compass that is depicted on the opposite page is to determine the direction of qi flow as well as the positioning of structures.

Finding Nature's Alignment

The concept of space, in terms of physical position and direction, has long been a potent force in Chinese culture. It is expressed in the body movements of t'ai chi (left), whose dancers seek alignment with nature's energy. It is even more evident in an ancient mystic art called feng shui, which for thousands of years has dictated the placement of Chinese homes, shops, palaces, tombs, and temples.

Feng shui, literally meaning "wind" and "water," holds that structures that harmonize with nature's alignment bring good luck; those that disrupt the flow of nature's animating force, or *qi,* are considered unlucky. Some feng shui experts claim to sense instinctively where qi flows smoothly; others determine auspicious sites with a geomancer's compass such as the one above, ringed by astrological and geographical details and symbols from the *I Ching,* the ancient text that summarizes the wisdom of the Orient. Indoors, feng shui experts hang mirrors in strategic locations to direct the path of qi. Or they may rearrange furniture or block off doors to balance a room's energy. Some say even the shape of a bed can affect the flow of qi and thus shape a person's destiny.

Officially, feng shui is repressed today in the People's Republic of China. Yet it is still surreptitiously practiced. And feng shui consultants enjoy a bustling business in Hong Kong, where their advice is sought not only by apartment dwellers and shopkeepers but by a number of international corporations that are headquartered there.

that the end of a cosmic cycle was due in that year. When the European invaders arrived, the Maya regarded them as a sign that the world was ending and fled without resisting.

The Maya and Nahua worshiped space with the same reverence as they did time. In Nahuatl geography, for example, each of the four compass points was assigned a sacred color. As the source of light, the east was yellow; in the south, which was death's home, everything was blood red. Like Hindus and Jains, the Maya believed in layers of heaven and hell, rising and descending from the earthly plane. Their thirteen layers of heaven revealed some of Mayan astronomy's limitations: Both the planet Venus and birds in flight were thought to occupy the same altitude, with the Milky Way located below them, and the moon and the clouds sharing the skies nearest to the ground.

Human souls in the Mayan cosmos were reincarnated and could, with supreme moral effort, rise to the highest heaven, the House of the Sun, where the supreme male-female god Ometeotl revealed to them the full nature of existence. Most souls, however, preferred to cycle endlessly through Tlalocan, the lowest heaven, where they passed the time feasting and playing games before returning to another mortal life again.

Almost all early cultures, including the Maya, believed in a sacred time, usually associated with the mythical period before time began. The Maya called it the first time, although to them it was actually a place outside time, the place where their god was before he "sounded the first word" that made eternity shudder and started creation. The Australian Aborigines called it Dreamtime and contrasted it with the worldly time in which they lived their day-to-day lives, what anthropologists refer to as profane time.

The Aborigines considered the occupations of profane time—hunting, building huts, trading, making love—as meaningless when compared with what they experienced in Dreamtime, also called Sacred Time. Fortunately, they believed they could visit this more desirable time by invoking it through religious ceremonies. For these rituals they decorated themselves, carried sacred objects, and chanted verses about their gods while venerating totem poles bearing the gods' images. Tribespeople spent as much of their lives as possible in Sacred Time. Since in Sacred Time, the birth of the universe was reenacted again and again,

Designed by the Mixtec of southwestern Mexico, probably in the late fourteenth century, this turquoise mosaic breast ornament of a two-headed serpent symbolizes time's dual nature—simultaneously creative and destructive, the bearer of both life and death.

participants believed they joined in the creation of the world, emerging from these ceremonies purified and strong. Likewise, whenever Aboriginal physicians administered a medicine, they first chanted the myth of the medicine's origin to make it more potent.

Although they are few and far between, there are still Australian Aborigines who live by these ancient beliefs. Every year, the men of the Arunta tribe reenact the story of their original ancestor. They prepare themselves by first enduring a period in which they fast and abstain from sex and fighting. Then, chanting their mystic verses, they immerse themselves in Dreamtime, where the long journey of the first Arunta occurs over and over again. They believe they are thus united with this godlike figure, and they emerge reborn and spiritually refreshed.

Similarly, certain Native American communities perform religious ceremonies that they believe carry their participants out of everyday time into a sacred timelessness. Their experience leads them to equate the idea of time with the idea of space. That is why the Yokut Indians of California say at the end of each year, ''the world has gone by.''

As advancing civilizations left their ancient roots behind, they grew nostalgic for the bygone be-

Sacrificing Lives for Time

Before the world began, according to Aztec myth, the gods gathered to decide who among them should light the new universe. They chose Nanautzin, a small, ugly, scab-covered god, who humbly accepted the task. When the time came, Nanautzin bravely hurled himself into a huge fire the gods had built. As his body crackled and burned, the flames leaped higher and higher, lighting up the sky and giving life to the sun. Thus with the willing sacrifice of Nanautzin, the cosmos was set in motion. But that sacrifice was not enough, ancient Aztecs believed, to keep the sun moving across the sky, to prevent time from coming to an end. More human sacrifices were required: Priests accordingly cut the hearts from living victims and, like the figure in the pre-Columbian bas-relief at left, offered them to the sun.

Although few carried human sacrifice to the grisly extremes of the Aztecs, many ancient cultures, including the Maya, Chinese, Greeks, Hindus, and Vikings, practiced it at one time or another. Eventually, however, the idea of sacrificing a life in order to honor a god or to maintain or renew life on earth began to be interpreted symbolically or mystically instead of literally. In the Old Testament, sacrifice is a test of faith: God demands that Abraham slay his son but stays his hand at the last moment. Christians believe Jesus Christ died for all humankind, a sacrifice that promises eternal life. Buddhism, on the other hand, totally rejects the idea of sacrificial killing, instead emphasizing that one must sacrifice the claims of the ego—all fears, desires, and self-interests—to attain enlightenment and honor the divine.

liefs that allowed them to escape the relentless march of profane time, which so often seemed to lead to a degraded and unhappy existence. But having lost the ability to easily transport themselves from profane time to a sacred time, they sought other answers, in new explanations for the universe, to replace the ancient, organic, cyclic mythologies of time and space they had outgrown.

The word *philosophy,* which means the systematic, rational search for order in the world and in human affairs, has its roots in the ancient Greek phrase meaning "love of wisdom." The Greeks were moved to originate this kind of intellectual inquiry by their desire to know the truth about time and space.

Like other ancient cultures, the Greeks viewed time cyclically. Their Great Year was a universal circle that revolved through the "gold, silver, bronze and iron ages" of civilization until all life was destroyed by a Great Winter and a Great Summer, eventually to be re-created and rolled through the ages of history once more. From this cosmology emerged the Greek faith in "eternal recurrence": Every event that had ever occurred, and every life that had ever been lived, would take place over and over again.

As traders who traveled throughout the Mediterranean, encountering diverse cultures and conflicting theologies, the ancient Greeks found themselves questioning their own religious assumptions. Eventually, they defied divine wisdom and searched for their answers solely through intellectual effort. If the gods ruled the mythic universe of the ancients, the mortal mind ruled the universe of the Greek philosophers.

Whereas in ancient mythologies time and space were divine entities inextricably entwined with life, the Greek philosophers gradually came to see time and space as abstract principles, separate from the material world. This led to some unusual notions. The early philosopher Parmenides, who recorded his thoughts in long poems during the sixth and fifth centuries BC, insisted that only things that were permanent and changeless could be real. Since time—which flowed from the past through the present to the future—was always changing, it must therefore not be real. And if time was not real, there could be no such thing as motion in space, which would necessarily involve time.

Parmenides' disciple, Zeno of Elea, brought these rarefied concepts down to earth in a series of paradoxes involving time and space, some of which continue to stump philosophers, or at least schoolchildren, 2,500 years after Zeno proposed them. In one example, Zeno "proved" that a flying arrow

was really not moving at all. He explained himself by challenging anybody to tell him how fast the imaginary arrow was flying at a specific point during its trajectory. One would have to admit, he said, that during any timeless instant, the arrow was not moving. It was simply filling up a

EVCLIDES

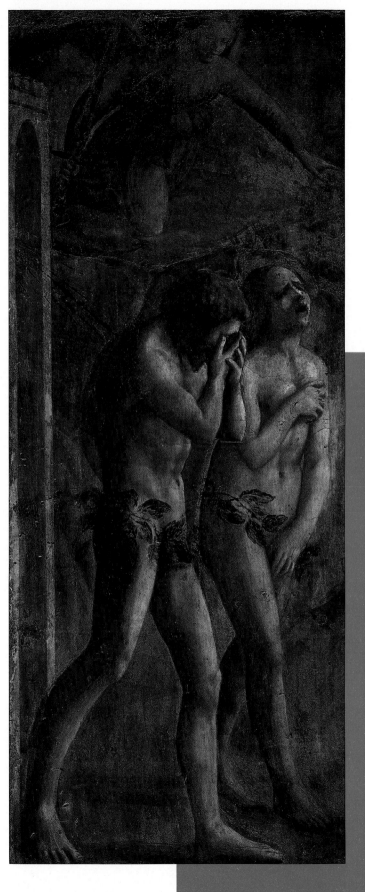

region of space equal to its length. Since the arrow was motionless, its apparent trajectory was but a trick of the observer's senses.

In another paradox, Zeno contended that it was impossible for a runner to cross the finish line of a race, no matter how fast he ran. In order to cover the distance to the finish line, Zeno argued, the runner would first have to get halfway there. Then he would have to run half the distance between the halfway mark and the finish, then half of that half distance. Since each time the runner would cover only half the distance remaining, he would never finish. "There are an infinite number of points in any given space," stated Zeno, "and you cannot traverse an infinite number of points in a finite time."

Not all of Zeno's contemporaries bought his arguments. When Diogenes the Cynic heard the case for Zeno's "racecourse," he got up and walked away to prove that movement was possible. But Zeno remained convinced that he carried the logical day. Time and space appeared to move, he said, because humans moved through time and space. In reality these phenomena were eternal and changeless.

The great philosopher Plato, who was born in 427 BC, some thirty years after Zeno's death, basically agreed with his Greek philosophical forebears, but he took on the task of discerning the nature of the reality behind these illusions of change. His brilliant efforts have led scholars ever since to say that "all of Western philosophy is but a footnote to Plato."

Plato argued that all of the objects of our senses—the world of space and time as we know it—are only imperfect, impermanent examples of pure forms that reside in a world of their own outside time and space. According to Plato, a line drawn on a page, even with a straightedge, is never perfectly straight or one-dimensional (since the pencil mark on the paper has some thickness). The only real line exists as an idea or an essence, from which all the lines humans make and see are imperfect copies.

Wracked with shame and despair, Adam and Eve are expelled from the Garden of Eden by the angel of God in this fifteenth-century fresco by Tommaso Masaccio. While Christians consider the creation the beginning of time, the couple's eviction marks the start of the human experience of time.

According to traditional Christian belief, just as God created time and space, he will end it with the Last Judgment, depicted in this detail from the Seven Deadly Sins, a fifteenth-century painting by Hieronymus Bosch. On that final day, all beings will be directed to the timeless realms of heaven or hell.

Our mundane world is imperfect, said Plato, because even God could not create something as perfect as himself. Instead, he made "an everlasting likeness" of eternity, revolving with numerical accuracy to arrange and define time. Time as we know it was thus tied to the physical world and would end if the earth ended. Plato believed every person carried innate knowledge of the other, timeless world of forms in "memories" from before birth, since each soul resided in that perfect realm before descending into a human lifetime. Through what he termed the dialectic, a process of rational self-inquiry, one could break the chains of illusion and rediscover this world of truth. Plato once demonstrated this universal human capacity by leading a supposedly ignorant slave through a dialectic interchange, until the slave was able to formulate a classic geometric proof: determining what the length of a square's side must be if the square's area is twice that of a given square.

Through all these intellectual breakthroughs, Plato established the supremacy of the human mind. Quoting his mentor Socrates, Plato proclaimed that "reason is king of heaven and earth." He even went so far as to say that good is nothing more nor less than knowledge; and evil equates to ignorance. By separating matter from spirit and elevating reason to a new stature, Plato and his compatriots laid the foundation upon which scientific investigations of time and space would be erected.

It was left to Saint Augustine, head of the Roman Christian church in northern Africa during the fourth century, to fuse the Platonic tradition with New Testament doctrine, creating the Greco-Christian concept of time that survives in the West to this day. Augustine believed, as did Plato, that when "God created heaven and earth," he also created time and space. God therefore existed outside time and space, and so would human beings in the form of immortal souls after the destruction of the universe. Augustine observed that earthbound human beings had a deep, intuitive sense of time passing, which was based on their memories and expectations, but could not conceptualize it. "What then, is time?" he asked. "If no one asks me, I know. If I wish to explain it to someone who asks, I know it not."

He made an attempt to explain it, however, basing his description on Judeo-Christian beliefs. God had created the heaven and earth "in the beginning." Thus time clearly had begun; it had not existed forever. "The world and time had both one beginning," Augustine declared. "The world was made, not in time, but simultaneously with time." And there would be an end to time, at least worldly time, when Christ returned to raise the dead.

In this manner, Augustine gave expression to the

Western concept of linear time, in which all things have a beginning and progress to an end, as opposed to the cyclic time of endless renewals embraced by so many other cultures. Within the framework of cyclic time, the past and the future always exist, along with the present. In Augustine's concept of time, only the present actually exists at any given instant; the past and the future are solely in the mind. Time is linear, a sort of ruler on which all events can be marked in straight progression.

The importance Augustine attached to the present conformed to New Testament teachings. Jesus was a historical figure, who accomplished God's work in a very short lifetime. Since time on earth was linear, and limited, how people spent their time was critical. There was literally no time like the present. Following Augustine's reasoning, the Christian church in the Middle Ages went so far as to cite belief in cyclic time as heretical. In a practical sense, traces of cyclic time persisted. Easter, for example, celebrated the cyclic-minded pagans' idea of the endlessly repeated return of spring, as well as the once-only resurrection of Christ. By tying human experience to linear time, however, Augustine banished from Western culture any notion that people had unlimited amounts of time to achieve spiritual perfection.

Over the centuries, other philosophers stepped forward to amend the world-view that Augustine willed to Western civilization. René Descartes, the seventeenth-century mathematician-scientist hailed as the father of modern philosophy, stated that space and matter were one and the same. What people thought of as distinct objects separated by empty space were, according to Descartes, integral parts of an indivisible reality that "stretched" like taffy to assume various seemingly distinct shapes.

Germany's Gottfried Wilhelm Leibniz, who was born just four years before Descartes died, came to another conclusion altogether—that there was no such thing as space at all. What people considered to be space, said Leibniz, was not a thing or substance but a system of relationships between objects. Employing the same kind of semantic argument, Leibniz claimed there was no such thing as time ei-ther, since time was merely a relationship between events.

In the next generation, the German philosopher Immanuel Kant went Leibniz one further, demonstrating, with arguments that a layperson might be tempted to label sophistry, both that time had a beginning and that time had no beginning. To prove that time had a beginning, Kant first considered the possibility that time instead was infinite—that time consisted of "an infinite series of successive states." If the past was indeed an infinite series of successive states, he reasoned, it could not have been completed in finite time; therefore, we could not be in the present. Since we know that we are in the present, time cannot have been an infinite series of successive states. If time is not infinite, it must be finite; it must have had a beginning.

His argument that time did not have a beginning was more straightforward. If time had a beginning, Kant said, that beginning must have been preceded by empty time. But empty time is time nonetheless; therefore, time could not have had a beginning. (A first-century-AD Roman philosopher named Lucretius had used the same basic argument to contend that space was infinite: If space had a boundary, what could there be beyond the boundary but more space? Therefore, space had no end.)

 hen not engaged in metaphysical debate, Kant had a realistically mundane understanding of the average human's grasp on the concepts of time and space. He suggested that space was our own invented idea of where things resided. He further concluded that our inborn sense of time would not allow us to comprehend an external, objective time, even if such a thing existed. Questions about the true nature of time and space, said Kant, were beyond human understanding.

As vigorously as European metaphysicians argued, people could not swallow the idea that the time and space they lived in every day were simply figments of their imagination. In the end it would become the task of science, rather than philosophy, to explain these age-old mysteries in a way that modern society could accept and live with.

Where Myth Met Science

Ancient peoples tried to understand space, time, and the universe through religion and myth; now we view those concepts through scientific eyes. Although the change was gradual, one man stood at the juncture where myth shaded into intellectual analysis, where ideas joined mysticism and science—a sixth-century-BC Greek named Pythagoras.

Pythagoras is best known for his familiar geometric theorem: that the square of the hypotenuse of a right triangle equals the sum of the squares of the other two sides. But more than a mathematician, he was a mystic who, some said, had magical powers; a musician who discovered the physics of the harmonic scale; and the first to call himself a philosopher, or lover of knowledge. He combined science and metaphysics in an idealistic blend perhaps best exemplified by his use of the word *kosmos*—which until then meant order or beauty—as a name for the universe. His followers, called Pythagoreans, said the cosmos was a harmony of balanced opposites such as odd and even, male and female, and good and evil. But the primal opposites were limit and unlimit—that which had form and that which was chaos. Pythagoreans believed limit acted upon unlimit to produce the mystical One, revered as the source of all numbers. And to them, numbers were everything.

A Universe of Divine Numbers

To Pythagoras, numbers were not an invention but were independent, holy entities, with traits to be learned through contemplation. One, the unit, was the primary source of existence. One and the next nine integers made up the sacred decad and were gods.

Pythagoreans studied only whole, positive numbers. They knew about fractions, negative numbers, and zero but shunned them as evil and impious. They depicted numbers not by symbols but by quantities, with rows of dots or pebbles like the array at right, which represents Ten. Though cumbersome for computation, this method revealed numerical traits that formed the basis of their number theory, a mixture of arithmetic and religion.

Any odd number, for instance, could assume an equal-armed L shape called a gnomon—literally, a carpenter's square—like the number Five, opposite. Odd numbers, or gnomons, were good, their squared shapes symbolizing regularity and equality. The even numbers, incapable of forming a balanced gnomon, were evil; they had oblong shapes, and oblong opposed the wholesome square in one of the basic dualities of the cosmos. Especially culpable was Two, which, like One, was considered not a number but a source of numbers—in Two's case, of all the even numbers.

Other esteemed numbers were those whose shapes were square, like Four (although it is even), or triangular, like Three or Six. Square numbers, as they are still called, result from multiplying a number, called the square root, times itself. The Pythagoreans noted that adding the appropriate gnomon to a square number yielded another square number. In this way, the gnomon Five fits along two sides of the square Four to produce another square, Nine—a process called gnomonic expansion.

Triangular numbers were revered as the sums of consecutive numbers, such as $1+2=3$. The Pythagoreans saw unique perfection in the triangular number Ten and worshiped it as the sum of divine power, adopting its triangular array *(left)* as an icon of their faith. This they named *tetractys*, or "fourness," because Ten is the sum of One, Two, Three, and Four, which themselves were considered to be the source of everything in the cosmos.

One, or Unity, was the essence of what Pythagoreans called limit. Although One is part of every number, it was defined as the basis of numbers and not as a number itself. It emerged from nothing, the Pythagoreans said, to give rise to everything else. One was identified with Zeus, father of the gods and creator of the cosmos.

Because One was the source of all the odd, or good, numbers—those that could be written in the salutary gnomon shape with equal arms—it was known as the friend, since friendship can occur only between equals. As the source of friendship and limit, One was the indispensable harmonizing influence in the cosmos, balancing all disparate elements into a whole universe. Since One was the agent and source of all existence, it was held to be hermaphroditic, both male and female. Similarly, while any number must be either even or odd, Pythagoreans said One was both.

Two, the Dyad, was female, material, even, and oblong. The opposite of the One in every way, Two was the essence of unlimit—chaotic, sprawling, and undisciplined. Like clay shaped by the sculptor's tool, Two acquired form only through the influence of One.

Among Two's many names was Daring, for its boldness in breaking away from the pure unity of the One. But its rebellion led to the evil and misery of material existence.

Three is the first digit that the Pythagoreans regarded as a number. Three meant plurality and multitude. It symbolized the world of matter, since only three points are needed to define a plane, or two-dimensional surface. Three was also the first number in sequence forming a triangle, which showed it to be the sum of successive numbers (1+2=3). It was said to have beginning, middle, and end; it stood for both the human psyche and the cosmic psyche that animated the universe.

Four is the first square number. Second in importance only to the One, Four signified justice. It made geometric solids—and all matter—possible, by adding the third dimension of depth. Only four points are required to define the vertices of the tetrahedron, the pyramidal solid figure that has as its sides four triangles. In signifying solidity, Four symbolized the creative urge that formed the cosmos, as well as the underlying numerical structure of existence.

Four completes the series needed to form Ten, the sacred decad (1+2+3+4=10), making Four itself another manifestation of the decad. This number series was divine and was the creative power in the universe. Many things in nature were said to occur in fours, including the cosmic elements themselves: earth, air, fire, and water.

Its position in the middle of the decad gave Five great significance. It symbolized marriage, because it was the sum of Three, the first male, or odd, number and Two, the first female, even integer. Thus it was sacred to the goddess Aphrodite. Five also signified the five regular geometric solids, whose faces are equilateral and equiangular: the tetrahedron, cube, octahedron (a shape with eight faces), dodecahedron (twelve faces), and icosahedron (twenty faces).

Six may be a triangular number (above left), the sum of successive integers (1+2+3=6). But when presented as an oblong number (above right), Six is seen as the product of two successive numbers: 2x3=6. From these properties come Six's distinction as the first perfect number: The numbers that divide evenly into Six (One, Two, and Three) also yield Six when added together. Perfect numbers were valued because they are rare; the next two are 28 (1+2+4+7+14) and 496 (1+2+4+8+16+31+62+124+248).

Six also represented the number of levels of living beings. The lowest were sperm or seed; then came plants, animals, humans, the daimones (who mediate between humans and gods), and at the highest level, the gods themselves.

Seven was called the Virgin, because it cannot be "generated" by multiplication of any pair of other numbers in the decad and because multiplying it by any number other than One produced a result outside the sacred decad. Seven multiplied by the sacred Four, however, yields twenty-eight, the second perfect number.

Eight, although it is an even number, was revered as the first cube number: 2x2x2=8. Furthermore, Eight equals 2+2+2+2, a combination so pleasing to the Pythagoreans that they named it Harmonia.

Nine is the first gnomon that is also a square number. Nine was sometimes called Oceanus—for the god of the sea—because, like the edge of the sea, Nine formed the boundary before the sacred Ten. Nine was also called Prometheus because, like that mighty god, it was strong enough to hold back the decad's other numbers.

Ten symbolized the ultimate and necessary good of limit and form, which stood in eternal opposition to unlimit and chaos. By interrupting the formless progression of infinity, Pythagoreans said, Ten made counting possible. Since all the preceding numbers of the decad were identified with gods, Ten meant the sum of the divine powers holding the cosmos together.

Harmony, Proportion, and the Golden Section

To their belief that number is the basis of all things, the Pythagoreans added another—that everything has its own proper proportion, which will result in harmony. In fact, they said, what underlies reality is not just number but an invisible web of ratios and proportions.

Pythagoreans valued ratio and proportion as hallmarks of reason; the very word *rational* meant the ability to comprehend ratios. Preaching and practicing moderation in all things, the Pythagoreans believed that the golden mean, or the middle way between extremes, was the path of wisdom. They counseled the middle way in marrying—neither above nor below oneself—as well as in food, drink, and exercise. In money matters, "never be prodigal," the sage advised, "nor show yourself stingy," but find the middle way.

In geometry, the middle way was the golden section, and it was esteemed as an ideal proportion. A golden section is a line so divided into two unequal parts that the smaller part is to the greater as the greater is to the whole; the lengths of the parts are said to be in the golden ratio. Not only is this division peculiarly satisfying to look at, but it occurs in nature. In working out their ideal proportions for human anatomy, the Pythagoreans declared that the navel should divide the human body into a vertical golden section. The golden section figures in some intriguing geometric operations, described and illustrated below. (Unfortunately for the Pythagoreans' predilection for whole numbers, the golden ratio cannot be expressed without using fractions. It is close to 5:8, but not exactly, and involves an irrational number, one that tails off in an endless string of digits on the far side of the decimal.)

Proportion also proved the key to another study that commanded Pythagoras's attention, music. He found that musical intervals—the difference in pitch between tones of the harmonic scale—depend on certain ratios in the lengths of strings tightened to the same tension. Happily, in these ratios the major notes were described using only the numbers in the sacred tetractys: One, Two, Three, and Four. The Pythagoreans even revised astronomy by describing the solar system as sun, moon, planets, and stars orbiting the spherical earth at fixed distances, which were in the same harmonic ratios as musical notes. These developments made the four primary numbers and their sum, the sacred tetractys, even holier and buttressed the disciples' faith that the entire cosmos could be described through whole numbers.

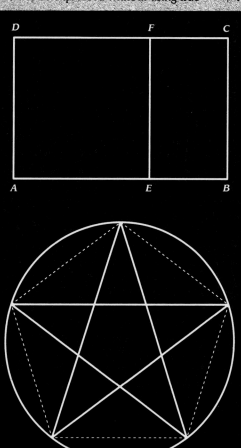

If unequal sides of a rectangle are in golden ratio proportion, as in figure ABCD at top right (the length of line AD is in the same proportion to the length of line AB, as AB is to the sum of lines AD and AB), the figure is a golden rectangle. Such a figure has a wonderful property. It can be subdivided into figures that remain in the same proportions, all partaking of the golden section. If a square (AEFD at right) is drawn to fill one end of the original rectangle, the remaining area (BCFE) is another golden rectangle, smaller than the first but identical in proportion. The process can be repeated infinitely, a fact that distressed the Pythagoreans, who double-damned infinity as "the Dark of Unlimit."

A five-pointed star, or pentagram, is formed by the simple trick of extending the sides of a regular pentagon until the lines meet. Alternatively, connecting the vertices of the same pentagon creates a pentagram inside the original figure.

Even beyond its ease of creation, what seems magical about such a figure is that every side of a pentagram cuts across two other sides at points that make them golden sections. Furthermore, the base of each triangular point of the pentagram is in golden ratio proportion with each of its sides. The Pythagoreans were so taken with these features that they named the pentagram Health and adopted it as a badge of their fraternal order.

The Golden Spiral and the Dark Secret

The Pythagoreans looked for evidence of divine harmony everywhere--in nature as well as in geometry. Among the richest expressions of harmony was the gnomonic expansion of numbers, whereby the appropriate gnomonic number, added to a square, produces another square number *(page 44)*.

Gnomonic growth creates the beautiful succession of curved compartments in the shell of the chambered nautilus *(left)*. Each sector is identical in shape to its neighbors; while they differ in size, all sectors have the same proportions. Moreover, they nestle within a golden spiral, the single curve built on the gnomonic expansion of a golden rectangle, as seen in the drawing at right.

Found in some seedpods and animal horns as well as in shells, the golden spiral symbolizes and embodies the harmony, regularity, and equality prized by Pythagoreans. It is a profound image for the movement of time, containing as it does the record of the past, the conditions of the present, and a complete prediction of the shape of future growth.

But the golden spiral was built on the golden ratio, which was an irrational number and clashed with their belief that only positive, whole numbers were the building blocks of the universe. Their religious rejection of the irrational *(box, below)* kept them from a full investigation of it. Even so, they laid the groundwork for later discoveries of mathematical patterns in nature, many of them based on the golden ratio.

Gnomonic expansion of a golden rectangle, by addition of a square onto the figure's long side, produces a nested series of ever-larger golden rectangles (right). Diagonals of the two largest rectangles intersect in the innermost figure. Arcs drawn outward from the intersection through the consecutive squares form the expanding curve (dashed line) known as the golden spiral.

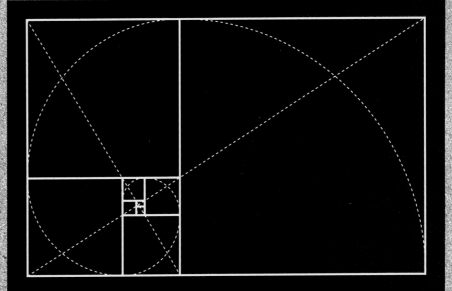

The Wrong Number in the Right Triangle

The Pythagorean theorem led the Pythagoreans directly to trouble. The theorem states that the square of a right triangle's hypotenuse (the side opposite the right angle) is equal to the sum of the squares of the other two sides. When both of the shorter sides of such a triangle measure one, the sum of their squares is two *(left)*. The length of the hypotenuse is the number, which multiplied by itself, gives two--that is,

the square root of two. But the square root of two is an irrational number, one that can never be exactly determined because the digits beyond the decimal point trail on ad infinitum.

The existence of such numbers challenged their belief in the primacy of whole numbers--so they kept it secret. They lumped irrational numbers with infinity and unlimit, relegating all three to the domain of "falsehood and envy."

1

1

square root of 2

CHAPTER 2

The New Visions of Science

he skies over Príncipe were sullen and overcast on May 29, 1919. Gloomier still were the spirits of the two British scientists—astronomer Arthur Eddington of Cambridge University and Frank Dyson, the British Astronomer Royal—who had traveled to this dot in the Gulf of Guinea to photograph stars during a total eclipse of the sun. For them, the cloud cover threatened to spoil everything. If they could capture clear images of the distant constellation Taurus while it was briefly visible during the eclipse, they could confirm or refute German physicist Albert Einstein's general theory of relativity. Up to that time, the theory had never been proved.

Among other things, Einstein's equations stated that light does not always travel in straight lines—as logic would seem to demand—but that it could be deflected as it passed close to a celestial body such as the sun. Eddington and Dyson planned to test this hypothesis by photographing the light of a group of stars in the constellation Taurus, which normally would not be distinguishable when passing close to the sun. Only the eclipse could make photographs possible. Analysis of the images would then reveal whether the light had been deflected or not. The morning had been rainy, however, and now, as the eclipse began, it appeared that their answers would be hidden by a layer of cloud.

Despairing of success, the pair dutifully set up their telescopes, and lo and behold, just as the eclipse neared totality, the scudding clouds parted long enough for two good photographs. Eddington immediately set to work analyzing the images, and after three days of feverishly calculating the positions of the stars in Taurus relative to the sun and earth, he knew that he had proof of Einstein's theory. The light from the stars in that distant constellation had bent under the influence of the sun.

Although Einstein later claimed that he had been so sure of the soundness of his theory that he had slept through the time of the eclipse, other scientists around the world were anything but indifferent. They realized that if Einstein's predictions were confirmed, all of science would have to come to terms with a radically new understanding of time and space. When the results of the Eddington-Dyson observations were announced before the

Royal Astronomical Society along with corroborative evidence from a second group that had journeyed to Sobral, Brazil, the story made headlines around the world. The twentieth century rushed to embrace its consummate genius, even though few people on the planet were able to comprehend his ideas.

Einstein's notions of relativity posed mysteries of space and time never dreamed of by the mythmakers of bygone ages or by philosophers up through the nineteenth century. Observing, experimenting, and conjecturing, scientists have since pursued the secrets of nature to the farthest boundaries of space. They have given us startling new visions of the cosmos every bit as profound and—in their own way as fantastical—as the creation myths of China, India, and the Mayan lands.

Cosmologists and astrophysicists have tempted us with prophecies of journeys into space and described the wonders we might find there—from galaxy clusters to black holes and neutron stars. Science has also given birth to a companion theory to relativity called quantum mechanics, which explains the workings of the universe on the very smallest scale—in the domain of electrons and other subatomic particles—in ways that Einstein's general relativity cannot. Many puzzles remain, but query by query, test by test, scientists are finally filling in gaps in our knowledge that have existed since the first human beings looked upon the sky in wonder.

Einstein's work was not, of course, the beginning of scientific inquiry into questions of time and space. While mythmakers and philosophers pondered the mysteries of creation, astronomers from as early as the thirtieth century BC had been trying to make sense of the physical evidence on display in the heavens. To nearly all early astronomers, it appeared that the earth was at rest in the center of a relatively limited cosmos and that the sun and the planets revolved around our world. The stars, they believed, were fixed on a distant sphere that surrounded the other heavenly bodies and the earth. In the second century AD, the Greek astronomer Ptolemy described an intricate and amazingly accurate model of the solar system based on such geocentric assumptions, which had been passed down through generations of scholars in Alexandria, Egypt.

During the Renaissance, however, this view of the cosmos slowly began to crumble. In the sixteenth century, the Polish scientist Nicholas Copernicus, who is often called the founder of modern astronomy, concluded that the earth circles around the sun, along with the other planets. Because such an idea ran counter to the teachings of the Roman Catholic church, Copernicus refrained from publishing his theory of a sun-centered cosmos until 1543, when he was on his deathbed.

Half a century later in Italy, Galileo Galilei constructed the first complete astronomical telescope, basing his design on reports of Dutch magnifying instruments. The telescope's magnifying strength was thirty-two power—about four times that of modern binoculars. Armed with this new ability to peer across the heavens, he was one of the first men to look upon the mountains of the moon and to see the spots on the sun. He discovered that the cloudy haze of the Milky Way actually consisted of countless individual

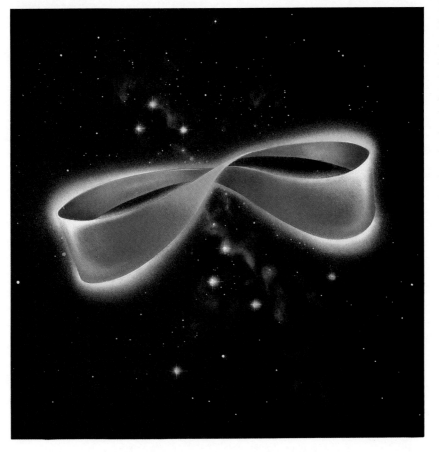

TABVLA III. ORBIVM PLANETARVM DIMENSIONES, ET DISTANTIAS PER QVINQVE REGVLARIA CORPORA GEOMETRICA EXHIBENS.

ILLVSTRISS: PRINCIPI, AC DÑO, DÑO, FRIDERICO, DVCI WIRTENBERGICO, ET TECCIO, COMITI MONTIS BELGARVM, ETC. CONSECRATA.

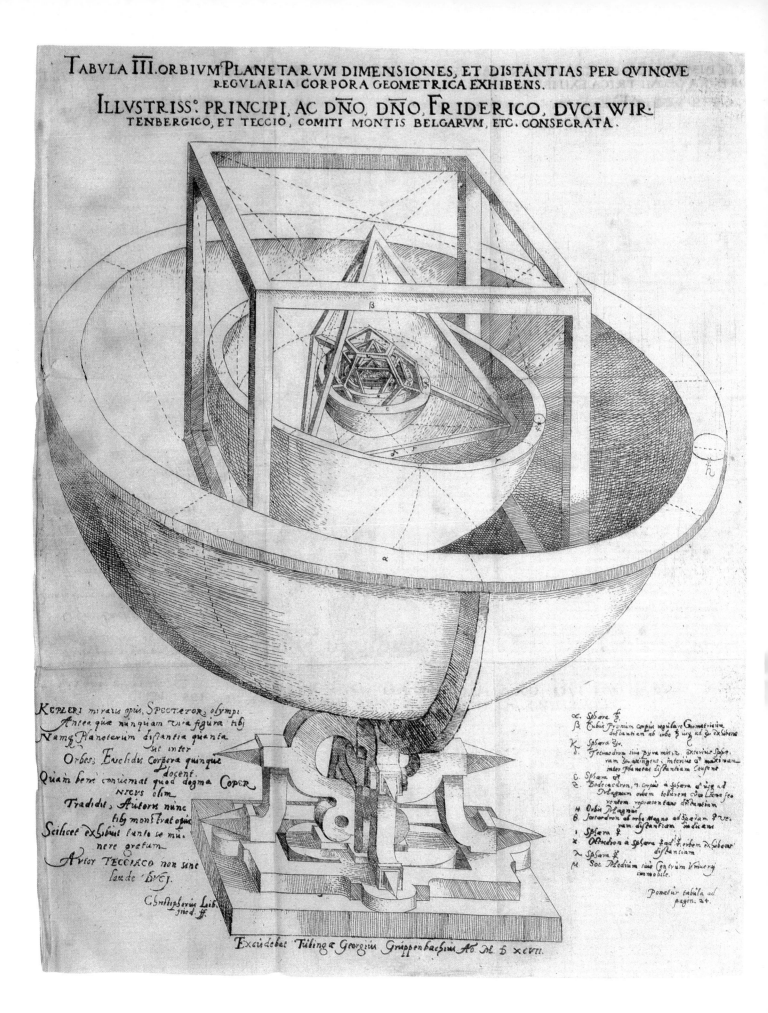

KEPLERI mirans opus, SPECTATOR, olympi.
 Antea quã nunquam vsa figura tibi,
Namq́ Planetarum distantia quanta
  sit inter
Orbes, Euclidis corpora quinque
  docent.
Quàm bene conueniat quod dogma COPER
  NICVS olim
 Tradidit, Autoris nunc
  tibi monstrat opus.
Scilicet exhibui tanto se mu-
  nere gratum
Autor TECCIACO non sine
  laude svej.

Christophorus Leib-
  fried. ff.

α. Sphæra ♄
β. Cubus Primum corpus regulare Geometricum
   distantiam ab orbe ♄ usq ad ♃ exhibens
χ. Sphæra ♃
δ. Tetraedron siue pyramis, 2. exterius Sphæram ♃ attingens, interius ♂ maximam
   inter Planetas distantiam consens
ε. Sphæra ♂
ζ. Dodecaedron, 3 corpus à sphæra ♂ usq ad
   Magnum orbem tollens cum Lunæ fe-
   rentem repræsentans distantiam
η. Orbis Magnus
θ. Icosaedron ab orbe Magno ad Sphæram ♀ ve-
   ram distantiam indicans
ι. Sphæra ♀
κ. Octaedron à sphæra ♀ ad ☿ orbem exhibens
λ. Sphæra ☿   distantiam
μ. Sol Medium siue Centrum Vniuersi
   immobile.

Ponatur tabula ad
   pagin. 24.

Excudebat Tübingæ Georgius Gruppenbachius Aõ M. D. xcvii.

stars, and he detected the moons of Jupiter and tracked their orbits. These and other observations convinced Galileo that Copernicus had been correct—the earth *did* move around the sun. Despite warnings from the Church, he made his beliefs known in a groundbreaking work called *Dialogue on the Great World Systems,* published in 1632.

Unfortunately, Italy was not ready for such a profound change in scientific outlook. Old-school professors denounced Galileo's radical ideas, while the Church declared them heretical. Threatened with excommunication or imprisonment, Galileo bowed to the authority of the Vatican; he publicly disclaimed his belief in a heliocentric universe. In private, however, he clung to his convictions and secretly authored another book, *Dialogues concerning Two New Sciences,* which he smuggled out of Italy for publication in Holland. This extraordinary attack on the teachings of the past has been cited as the very genesis of modern physics. Galileo died old and blind in 1642, but not before he had touched off a revolution in cosmological thought. The same year that Galileo expired near Florence, the next—and perhaps the greatest—pathfinder of science was born in Woolsthorpe, England.

Isaac Newton's vision embraced the entire universe. His theories of motion and gravitation defined the relationship between objects in space. In a brilliant mix of geometry and mathematical equations, he precisely described the period of the moon's orbit around the earth and the orbits of comets and planets around the sun. He even explained how gravity controls the tides upon the seas. Within the boundaries of the universe, he said, the motion of any object was relative to that of all others. But the framework of the universe itself was unchangeable, he believed. Newton referred to this framework as "absolute space," and in *Philosophiae Naturalis Principia Mathematica,* he wrote that "absolute space in its own nature without relation to anything external, remains always similar and immovable."

The *Principia* was Newton's masterpiece. It would shape all scientific thought on the workings of the universe for more than two centuries to come.

In Newton's view, time was unfailingly steady and reliable, like the framework of space. It flowed evenly and "without relation to anything external." Newton believed that time and space were constant verities against which all else could be measured.

There were, however, several troublesome theoretical questions left unanswered by Newton's grand scheme. One concerned the workings of gravity. How could this force, which holds the universe together in exquisite balance, be applied instantaneously from a body in space to others that were great distances away? Equally vexing to the students of Isaac Newton was the lack of understanding of the mechanics of light traveling through space.

By the nineteenth century, physicists postulated that light traveled in waves. But how? Sound waves, the physicists knew, need a medium through which to travel, such as water or air. They could prove this by placing a clock in a bell jar, then drawing all the air from the jar. In the resulting

Sir Isaac Newton described a universe in which time and space were absolute constants and events proceeded as in some giant, intricate clock.

*German mathematician Hermann Minkowski championed young
Albert Einstein's theories of relativity. "Space by itself," he told his colleagues,
"and time by itself, are destined to sink completely into shadows"
and give way to "a kind of union of both."*

vacuum, the clock could not be heard. Yet the clock was still visible inside the jar, so air clearly could not be the medium of light waves. To explain this enigma, physicists theorized the existence of an altogether different medium, which they called ether. Weightless, invisible, and stationary, this mysterious substance was thought to permeate the universe and all its many parts. It transmitted light waves but was itself motionless, a component of Newton's absolute space.

The notion of ether was a neat solution and widely subscribed to by scholars. But not many scientists were entirely comfortable with a theoretical proposition that could not be tested satisfactorily, and the existence of ether seemed impervious to experimental proof. Still, nothing else seemed to explain the movement of light, so the existence of ether endured as a theory until the eve of the twentieth century.

In 1887, a German-American physicist named Albert Michelson unwittingly showed that ether could not possibly fill the vacuum of space. Actually, Michelson was trying to find out how fast the earth moves through space (a speed that has since been determined to be about eighteen miles per second). He reasoned that as the earth moves through the ether it must create a sort of windlike disturbance, which he called the ether drift. This wind, he assumed, would in turn affect the velocity of light passing through it. Thus, if Michelson could measure the effect of ether on the speed of light, he could calculate the earth's velocity.

With his partner, chemist Edward Morley, Michelson designed an ingenious device that beamed two rays of light at right angles to each other over precisely equivalent distances and reflected them back to a timing device. One light beam was sent in the direction of the supposed ether drift; the other was pointed across it. Michelson predicted a minuscule but measurable discrepancy in the speeds of the two beams. The difference, he reasoned, would show the velocity of the earth's ether drift.

To his dismay there was no difference whatsoever. Refining their equipment, Michelson and Morley repeated their experiment again and again, always with the same result. The light beams traveled at precisely the same speed no matter what their direction relative to the earth's motion. Michelson refused to believe the evidence of his tests and abandoned the project as a failure.

It would take a twenty-six-year-old amateur physicist, temporarily employed as a patent inspector, to make sense of the Michelson-Morley experiment. At the start of his career, Albert Einstein had impressed few around him as destined for outstanding achievement. But already he had begun to create for himself a stunning new framework for exploring the mysteries of time and space—the approach that came to be called relativity.

The familiar countenance of Albert Einstein in his later years—a disheveled, wild-haired icon of genius—is still as identifiable decades after his death as that of any president or movie star. His name became a synonym for piercing intelligence, and he spent his last twenty-two years at the Institute for Advanced Study in Princeton, New Jersey, revered as a prototype of the Great Thinker.

Yet little in his early life marked him for greatness. He was born into a middle-class Jewish family in Germany and was a slow learner—verbally, at least. As he recalled later in life: "My parents were worried because I started to talk comparatively late, and they consulted the doctor because of it." But he never forgot his excitement at the age of four or five, when his father gave him a compass for a toy. He was fascinated that the swinging needle in its little case could respond to the unseeable force of magnetism.

At school, some of his teachers considered him dim-witted. Although he clearly had a bent toward equations and problem solving, his ineptitude at learning Greek led one teacher to warn him that he would "never amount to anything." Young Einstein left school before graduation and barely gained admittance into a technical college whose physics courses, he said, "had such a negative effect on me that after my finals the consideration of any scientific problems was distasteful to me for a whole year."

After graduating from college, he worked as a tutor and, in his spare time, dabbled in philosophy and music with his friends. (Einstein was a better-than-average amateur violinist, and music became for him a lifelong source of pleasure. Mozart was his ideal.) Yet despite the discouragements he had met so far in his academic career, Einstein apparently had a sure sense of his gifts for mathematics and science. In 1902, he landed a job as technical expert, third class, at the patent office in Bern, Switzerland. He devoted his free time to physics.

One of the first tasks Einstein undertook was an analysis of the physical properties of light: A German physicist named Max Planck, who was to become one of the pivotal figures in the development of quantum theory, had recently announced his belief that atoms radiate energy not in waves but in particles. Einstein realized that if light, which is a form of radiation, consisted of particles, it would not need a medium to convey it. Based on this one intuitive leap, he concluded that ether simply did not exist. He was then able to take Michelson's results at face value and infer an entirely unexpected conclusion—that the speed of light is unaffected by the relative movement of bodies in space.

This thesis directly contradicted Newton's laws of relative motion. Newtonian physics dictates that the relative speed of two objects in motion is additive. For example, if a quarterback throws a football downfield at a velocity of 30 miles per hour and the receiver is rushing downfield at 15 MPH, the ball will float into the receiver's arms at a gentle 15 MPH. A defender racing at the same speed in the opposite direction in hopes of intercepting the pass, though, would have to cope with the ball traveling at a relative speed of 45 MPH. This way of calculating relative motion is confirmed by many other common experiences as well.

Einstein, however, argued for uncommon sense, insisting that light behaves differently. No matter how fast the source of light is moving, he suggested, and no matter how fast or in what direction any observers are traveling, the speed of the light will always be the same. If the velocity of light is a constant, Einstein theorized, then all motion and even time itself—by which motion is measured—must be relative to it. With this bold stroke, Einstein did away with Newton's notion of time as

The Relative Movement of Time

3:00

3:00

3:00

Time and space are not absolute and independent, but variable and interconnected, as Einstein proved in his special theory of relativity. Because of the interrelationship of space and time, the faster a person moves through space, the more slowly time passes. This is best shown by a hypothetical interplanetary voyage. A spaceship approaches two planets. On each planet is a huge clock that has been synchronized with earth time and shows 3:00. Passing the first planet, the astronauts see its clock and synchronize their ship's clock with it. They set the on-board clock to 3:00.

4:00

4:00

The spaceship is moving at 140,000 miles per second, or 75 percent of the speed of light. At this speed, the astronauts compute, they should reach the second planet, 504,000,000 miles away, in one hour. Passing that planet (right), they see that its clock, as expected, reads 4:00—but the ship's clock reads only 3:40. Why? Einstein called it "time dilation": The speed of their movement has dilated, or stretched, time. A clock on a moving spaceship will run slower than a stationary clock, and the faster the ship moves, the slower its clock will run.

3:40

an absolute. And he forced the world of science to rethink its perceptions of all time-measured functions, including speed, acceleration, momentum, and energy. If time is relative, so is everything connected with it—with the single exception of the speed of light.

Nothing in the universe, Einstein insisted, can travel faster than light, which moves through the vacuum of space at a velocity of 186,272 miles per second. This central thesis of Einstein's so-called special theory of relativity led to some surprising predictions. It suggested, for example, that if a human being or any other physical object could travel at a velocity close to the speed of light, then the age, mass, and size of that person or thing would be drastically different for anyone observing from a stationary vantage point. Time would dilate, or slow down. The faster a clock raced through space, the slower it would seem to tick to the stationary observer. At the speed of light itself—the clock would seem to stop altogether and time would stand still. In the same way, distances would appear to shrink: A yardstick projected at extraordinarily high speeds would get shorter as it traveled. And the mass of both the yardstick and the clock would seem to increase as their velocity increased. At the speed of light, their mass would reach infinity.

Albert Einstein, at twenty-six, published his theories of relativity while employed in the Swiss patent office—"that secular cloister," he later commented, "where I hatched my most beautiful ideas."

Obviously, such possibilities have little to do with the sluggish world of everyday life. But the changes predicted by Einstein's theory have been proved real nevertheless. Even at the comparatively leisurely speeds attained by the most advanced spacecraft today—the space shuttle orbits at about 17,000 MPH, or approximately $1/40,000$ the speed of light—an atomic clock carried aloft ticks more slowly than its counterparts back on terra firma. (The story is told that the ground crew for one shuttle flight threatened to dock a small fraction of its astronauts' pay, because the space travelers were putting in slightly shorter workdays than

their earthbound colleagues.) Scientists debate whether or not it will ever be possible to build spaceships capable of moving at even one-third the speed of light (approximately 200 million miles per hour). If such craft do someday carry humans out to the stars, the clocks on board would slow down only about three and a half minutes per hour.

In the particle accelerators used by physicists to study the makeup of atoms, however, the effects are less hypothetical. Subatomic particles can be driven through these giant devices to speeds more than nine-tenths the velocity of light. At that level, the changes predicted by the special relativity theory become obvious, with the mass of the streaking particles increasing just as Einstein calculated. Normally short-lived particles exist significantly longer at near light speed, proof that for them time has slowed down.

The 1905 scientific paper in which Einstein spelled out his special relativity theory was called "On the Electrodynamics of Moving Bodies." At the time it was published, he had not yet worked out the famous $E=mc^2$ equation that is forever linked with his work. Arguably the most widely known equation ever written, this elegant formula distilled Einstein's perception that energy and mass are equivalent and that either one can be transformed into the other.

This insight grew out of Einstein's calculations predicting an increase in the mass of an object at speeds close to the velocity of light—designated c in the equation. He concluded that the added mass of the object was derived from the energy of its motion. It took the great physicist another two years to realize that the reverse must also be true: All mass must have energy. His $E=mc^2$ equation was eventually published in 1907. The awesome truth it stated so succinctly was corroborated nearly four decades later by the atomic fireball that blossomed over the New Mexico desert at Alamogordo on July 16, 1945.

 instein's first theory of relativity dealt with the nature of light and the effect of uniform motion on the physical aspects of mass, energy, and time. He spent most of the next decade wrestling with an explanation of gravity. His goal was to resolve the shortcomings in Newton's description of this fundamental force and to incorporate gravity into his new relativistic picture of the universe. The task required almost superhuman cerebral effort, as Einstein struggled on the furthest frontiers of mathematics and theoretical physics. "In all my life I have never before labored at all as hard," he wrote when his brainchild, titled "The Foundations of the General Theory of Relativity," was published in 1916. The work would be the crowning achievement of his life.

General relativity is an immensely complex theory—one that is daunting even to many serious students of the sciences. When Arthur Eddington, who is renowned for his work in explaining relativity to the English-speaking world, was asked by a journalist whether it was true that only three people in the world understood the subject, he paused, then replied: "I'm trying to think who the third person is." But some of the ideas incorporated into Einstein's grand scheme have wielded an influence far beyond the limited sphere of scholars who have plumbed the intricate depths of relativity. One such idea is space-time.

In our everyday world, space has three measurements—height, length, and breadth. But in Einstein's universe, space and time are inseparable and together form the four-dimensional continuum that he called space-time. Since time is relative—or put another way, elastic—space-time is, too, and the continuum expands and contracts throughout a finite but unbounded cosmos. We cannot see this four-dimensional space-time or even conceptualize it very well. In the language of sophisticated mathematics, however, it makes perfect sense.

According to Einstein, the force that the stars and planets in the continuum exert upon one another is not as simple as mere attraction attributable to their gravity, as Newton saw it. Each body in the universe acts upon the shape of space-time itself, warping it into complex four-dimensional curves. Anything moving through a star's space warp thus follows the precise trajectory of the curve in that region of space. An orbiting planet, like the slow-moving earth, follows the curve in an endless ellipse. The 600-trillion-mile-an-hour starlight that Eddington triumphantly photographed on Príncipe in 1919 had dipped fleetingly into the warp of the sun and then continued on its altered course. In a similar manner, the extraordinarily successful *Voyager 2* space probe, launched in 1977, has pursued its photographic mission through the solar system on a course that has sent it swooping from warp to warp.

In 1929, American astronomer Edwin Hubble made a discovery that would demand as great a change in our understanding of space as had the theories of Einstein. In break-

The Irreversible Arrow of Time

Questions about the nature of time have intrigued philosophers and scientists for centuries. What is time? Does time have a beginning and an end? Can time go backward? Great thinkers such as Saint Augustine, Galileo, Sir Isaac Newton, and Albert Einstein devoted countless hours to these questions. Although a detailed study of time demands a working knowledge of calculus, theoretical physics, and other mathematical disciplines, concepts such as the "arrow of time" can help laypersons grasp the fundamentals of this complex subject.

The arrow of time concept demonstrates the constant movement of time, distinguishes the past from the future, and gives time a direction. And as the artist's rendering here suggests, as the arrow moves forward, orderly form decays into disorder and formlessness. First suggested by the twentieth-century British astronomer Sir Arthur Eddington, the idea is based on the physicists' second law of thermodynamics, which states that in any isolated system, en-tropy—or disorder—increases with time.

A classic example of an isolated system with a high degree of order is an ordinary drinking glass. If the glass falls and shatters, disorder increases. The reverse will never occur: The shards of glass will never leap into the air and reassemble themselves. Events move steadily and irreversibly forward into the future, never turning back into the past.

Therefore, explain scientists, time must act correspondingly, moving inexorably into the future. Although, as Einstein explained in the special theory of relativity, time moves slower or faster depending on the observer, it still moves in only one direction, and that is forward.

Scientists also speak of other arrows of time. Human memory, which can remember the past but not the future, provides the psychological arrow; and the universe itself, always expanding, provides the cosmological arrow. Each demonstrates that time continues its ineluctable flight into the future.

ing down light from very distant stars into its component colors, he found that the spectral lines amid the varied hues contained in the light were further toward the red end of the spectrum than he had expected. Perhaps more important, he discovered that the extent of this so-called redshift was in direct proportion to the distance of the light source from earth. Hubble's new evidence sounds inconsequential, but it confirmed the suspicions of some cosmologists that the universe is not static but expanding.

This was a jolting revelation for most astronomers. Up to that time, they had yet to adapt to Einstein's notion of flexible space-time and still believed, with Newton, that the universe was static. Even Einstein, who rarely hesitated to challenge old ideas, had refused to believe that the universe could be expanding. His own general relativity theory had predicted as much, but he had balked at such a foreign concept. Einstein had inserted into his equations a fudge factor, called a cosmological constant, which canceled out the mathematical indications that the matter in the universe should be flying off in all directions at stupendous speed. Einstein later admitted that this narrow-mindedness was the greatest mistake of his life.

The key to interpreting Hubble's discovery lay in the so-called Doppler effect, which is most familiar in its auditory manifestations. When a train rushes by, for example, the sound of the locomotive whistle descends in pitch. The whistle's sound waves are getting longer, relative to the listener, so the sound is lower. The redshift that Hubble noted was the same sort of "Doppler down" effect in the light waves reaching earth from distant galaxies. With the help of Hubble's new rule for interpreting redshift, astronomers discovered that, not only were other galaxies receding from our own in all directions, but the farther away they were, the faster they moved away. His findings, though later modified somewhat, have withstood a half century of testing and are now universally accepted by scientists.

The realization that the universe was expanding confronted scientists with a fundamental question: Where were all those innumerable galaxies expanding from? This mystery cuts to the heart of the enigma of creation and has prompted many theoretical solutions. So far, the answer that appeals most to the logic of scientists is the one universally and appropriately known as the theory of the big bang.

The concept of a colossal explosion in which the universe began expanding from a state of infinite density was first proposed in 1927 by a young Belgian priest named Georges Lemaître, who was also a brilliant scientist. If the universe was now moving apart, he conjectured, it must at one time have been smaller. Imagining the expansion in reverse, he envisioned galaxies rushing toward one another, coming ever closer together. Carrying the scenario to its logical conclusion, Lemaître pictured all matter and energy joined in what he called a primordial atom. This cosmos, shrunk beyond the ability of the eye to see, was conceptually little different from the cosmic eggs of ancient mythology. But the "egg" of the scientist was marked by unimaginable density and heat. Our universe began, Lemaître suggested, when a cataclysmic explosion spewed all matter and energy into its still unfolding condition.

By the 1940s, Lemaître's cosmic-egg theory had been dubbed the big bang. (Ironically, the nickname was first applied derisively by British astrophysicist Fred Hoyle, who did not accept Lemaître's idea of creation.) Since that time, however, physicists, astronomers, and cosmologists have stretched the laws of relativity to their limits in order to arrive at a wary general agreement about how the universe might have begun in a big bang—if indeed such an event ever occurred.

Lemaître's cosmic egg is now referred to technically as a "singularity." It is believed to have been a solitary point of infinite density, which had no width, depth, height, nor any other measurable dimension of space-time. No one is sure what caused this speck of everything-and-nothing to explode, but big bang theorists have worked out a possible timetable of events for the split second immediately following the blast. Some of their descriptions deal in increments of time that are so small they are almost unimaginable. For example, the emergence of specific atomic forces took place at 10^{-44} second after "ignition,"—that is, .00000000000000000000000000000000000000000001 of a second after the big bang began.

A fraction of a second later, the furnace of expanding energy had cooled enough to "freeze out" matter in the form of atomic particles. (Since the temperature was still billions of degrees centigrade, the concept of freezing is, of course, relative.) At this point, according to a theory proposed by American physicist Alan Guth, a brief but sensational "inflation" took place, during which the nascent universe doubled in size 150 times, to about the dimensions of a grapefruit. Approximately one ten-thousandth of a second had elapsed since the bang began.

From that point on, the time frames in the scientific descriptions become more imaginable. After 300,000 years, for instance, the universe is believed to have cooled enough for the particles to form into atoms and molecules of superheated gases, the raw materials of stars. In due course, the stars bunched into galaxies, where, it is supposed, small irregularities in the cosmic soup may have caused particularly strong pockets of gravitational attraction. Cosmologists estimate that even though between 10 and 20 billion years have passed since the time of the big bang, the galaxies and the space they inhabit are still racing away from their mother egg.

Theorists are divided on the question of where all this matter and space will eventually wind up. The answer depends on whether or not the total amount of matter in the universe exerts enough gravitational force to ultimately

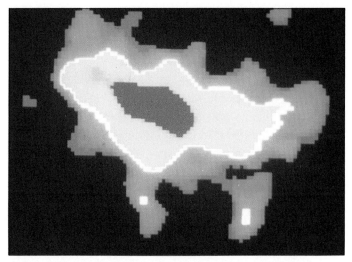

Galaxy 4C41.17, the most remote galaxy yet seen, appears in a computer-enhanced photograph that enables scientists to gauge its distance from earth. Detected only in 1988 by radio telescope, 4C41.17 is about 15 billion light-years away, near the edge of the visible universe. Viewed today as it was 15 billion years ago, 4C41.17 suggests that some galaxies formed just a few billion years after the big bang, much earlier in cosmic time than current theory allows.

slow down the expansion. If not, we exist in what the scientists call an open universe, which will keep on expanding forever—or at least until all the galaxies have burned out and the energy of the cosmic egg has dissipated into a frozen emptiness. If the amount of matter is sufficient, however, gravity will curve space-time back in on itself and the fleeing galaxies will gradually slow down. In that case, the time will come—though not less than 10 billion years from now—when the galaxies will begin to retrace their journey. The universe will contract until everything comes together again in the ultimate "big crunch." Ultimate, that is, unless the whole process begins again with another big bang.

Which will it be, endless expansion or the big crunch? British theoretical physicist Stephen Hawking voices the mainstream scientific opinion when he says: "I don't really claim to know the fate of the universe. Nor does anyone else. I think the best guess is that it is just on the borderline between collapse and expansion. But that is just a guess."

New telescopes, such as those being put into space beyond the earth's obscuring atmosphere, may eventually help scientists find out whether the universe has enough matter to cause its gravitational collapse. Then the fate of the universe may be clearer. And new research may tell us more about the causes of the big bang. But we will still be left with the question that has been asked since the dawn of human history: What was there before the beginning?

As in Newton's day, the answer for many people is found in religion. The Roman Catholic church, for instance, has officially accepted the big bang concept as compatible with the teachings of the Bible. For many scientists, however, the very question is moot. "If the suggestion that space-time is finite but unbounded is correct," says Hawking, "the big bang is rather like the North Pole of the earth. To ask what happens before the big bang is a bit like asking what happens on the surface of the earth one mile north of the North Pole. It's a meaningless question."

Stephen William Hawking belongs in the line of great European physicists—almost uncannily so. He was born in 1942, 300 years to the day after Galileo's death, and as Lucasian Professor of Mathematics at Cambridge, he holds the same academic chair that Newton held. Many scientists consider him to be the most brilliant theoretical physicist working today. As one associate put it, Hawking is "the nearest thing we have to a living Einstein."

Hawking's mental abilities were honed, perhaps, by extraordinary physical adversities. In his early twenties he was afflicted with amyotrophic lateral sclerosis, commonly known as Lou Gehrig's disease, and he has spent much of his life since that time confined to a wheelchair. He has been unable to move anything except his eyes and three fingers of his left hand, and in 1985, he lost the use of his voice. In spite of these obstacles, he has played an active and incisive role on the frontiers of physics theory. Since

Wheelchair-bound, all but completely para-lyzed, English physicist Stephen Hawking, the world's leading authority on black holes, devel-ops theories and formulas as complex as sym-phonies in his head, before expressing them through a voice synthesizer. His pioneering work at Cambridge University is aimed at knit-ting together relativity theory and quantum mechanics—the cosmic and the subatomic realms of modern physics—into a grand unified theory. Hawking explains his lifelong pursuit in terms that are at once sweeping and simple: "I want to understand why the universe exists at all," he says, "and why it is as it is."

losing his voice, he has relied on a computerized voice syn-thesizer to enable him to communicate.

Hawking has devoted his most intense efforts in re-cent years searching for an overarching theory that would unite the varied theoretical disciplines of physicists. If he succeeds in developing such a so-called grand unified the-ory, it will undoubtedly be remembered as his greatest leg-acy. In the meantime, he is best known for his contributions to the intriguing subject of black holes. These so-far unver-ified bodies in space are predicted by theory to be extremely compact but possessed of such great gravitational force that no radiation can escape them, not even light. They may prove to hold the key to our understanding of the big bang. Hawking's interest in black holes dates back to his days in graduate school, when he was exploring the intricacies of general relativity.

Collaborating with mathematician Roger Penrose, he proved that if Einstein's theories hold true down to the very smallest scale of things, then objects such as black holes simply must exist. He spent many years working to substan-tiate his belief that black holes are related to the phenome-non of the big bang. In the process, he seems to have con-vinced most of his colleagues that his thesis is correct.

The name *black hole* was applied to these mysterious occupants of space in 1969 by the American scientist John Wheeler, but the idea has been around for more than 200 years. In 1783, an English astronomer named John Michell suggested that a star could be so massive that the force of its own gravity would cause it to implode. In that eventual-ity, the gravity of the star would then be sufficiently intense to prevent any light rays from escaping. The resulting chunk of space wreckage would be invisible, but it would exert enormous gravitational influence upon the area of space around it. Hawking, for one, is convinced that these strange celestial bodies will be experimentally verified before long.

The first obstacle to finding a black hole, of course, is that it cannot be seen, since no light can escape its tremen-dous gravitational field. Nor will it be possible to catch a black hole in the process of formation. When a star collaps-es to form a black hole, it does so much faster than the eye can see. The light from a star ten times as massive as our sun, for example, would be extinguished within four-millionths of a second.

A more likely signpost to the existence of a black hole may be found in the space-time warp of its gravitational field. Discovering such a warp is theoretically possible, even though black holes, for all their potency, are but small dim-ples on the vast expanse of the universe. Scientists believe that if a black hole is relatively close to some other celestial body, the effects of its gravitation could be detectable.

Astronomers suspect they may have found precisely such a situation in the distant galaxy known as M-87. They have detected a huge jet of light reaching out from the gal-axy's core (a possible consequence of great amounts of matter spiraling into the galactic center), and their calcula-tions reveal that there is not enough visible mass in the gal-axy to account for the rapid motion of its stars. Both of these anomalies could be explained by the presence of a supermassive black hole, with a mass possibly 10 billion times greater than that of our sun. Another black-hole pros-pect—this one far less massive—is the two-star system called Cygnus X-1. One star of the pair is visible, while the other is not—possibly because no light can escape it.

lthough no certified black holes have yet been discovered, astrophysicists are rea-sonably sure they know how such objects are formed. A typical black hole of the kind suggested by Cygnus X-1 begins as a mas-sive star, at least five times as large as our sun. Without this great mass, the star would not have enough gravity to generate its own destruction. After con-suming itself in normal nuclear fusion for several billion years, the star begins to cool, and at a predictable point, it collapses under the force of its own gravity. (Some stars expand first in a fiery glow; they are called red giants.)

As the dying star shrinks, it reaches a critical point where gravity fights against the internal forces of atomic matter. If the star is not large enough to sustain the process, these atomic forces will bring a halt to the shrinkage, and the result will be either a white dwarf or what is called a neutron star. Our sun is destined to become a white dwarf in another billion years or so. It will end up slightly smaller than the earth, but with vastly greater density: A chunk the size of a matchbox might weigh in the neighborhood of ten tons. A neutron star winds up smaller still, perhaps no more than twelve miles in diameter. Its density is so great that a matchbox-size sample might weigh several billion tons.

But if the star is very massive to begin with, nothing can stop its slide into total oblivion. Just as the star finally implodes, the light that until then has been illuminating the violent decay of its matter becomes forever trapped behind a boundary of gravity that is known as the event horizon. A black hole is born.

 eneath the event horizon, all of the matter that once constituted the star becomes infinitely condensed so that it is left with no dimension whatsoever. It becomes, in other words, a singularity like the one at the heart of the big bang. Singularities are unlike anything we have actually observed in the physical universe. The laws of relativity lose their jurisdiction in these places of infinite density and extinguished time. There are no rules to describe what is happening or to predict what will happen next. This absence of order and logic may account for scientists' initial reluctance to accept the possibility that there were singularities in every black hole. But Stephen Hawking's equations proved this to be the case.

Black holes, neutron stars, red giants, and white dwarfs are all particles of cosmic space, where the serene laws of relativity hold sway. There is a whole other world of space and time, however, in which many of the rules seem to be different. In the microspace of the atom, the ruling code of law is quantum mechanics. And just as relativity reshaped our understanding of cosmic time and space, quantum mechanics has led to revolutionary discoveries about the nature of things in the subatomic domain.

In this realm—roughly as much smaller on a human scale as people are smaller than the sun—measurements are made in units on the order of one-quadrillionth of a meter. Here, elemental particles whiz about in a blur of motion, leaping from one state of being to another, losing and gaining energy, dematerializing and reappearing in different guises. This quantum ballet is apparently orchestrated with great precision by basic atomic forces, but the nature of its organizing principles is not yet fully understood.

For all the remaining mysteries, however, quantum mechanics has already proved to be a thoroughly practical discipline. Ideas spun off from this branch of science have filled our shopping malls with a bounty of high-tech electronic devices, from transistor radios to digital watches and computers. Likewise, these studies have led directly to breakthroughs in laser beams and nuclear power. At the theoretical level, quantum physics is drawing scientists ever deeper into the secret world of the big bang and—they hope—closer to answers about the fabric of the universe.

The quantum theory was born with the twentieth century. It dates from the year 1900, when Max Planck proposed the particle theory of atomic radiation that Einstein was to apply to his study of light. Planck had found that the wave theories that held force in his time did not satisfactorily explain the results of his own experiments. So he theorized that atoms discharge their energy not in waves but in discrete bunches. He called these energy packets quanta, from the Greco-Latin word for "how much." To this basic unit of energy he assigned the symbol h, now referred to as the Planck constant and considered as fundamental a fixture of nature as Einstein's c for the speed of light.

When Planck first published his theory, however, no one paid much attention. Although the evidence from Michelson's experiments had been available for a dozen years, the wave theory was too deeply rooted in the thinking of scientists to be easily dislodged. Even Planck was not entirely satisfied with his theory; he had come up with it, he later admitted, as an act of desperation, when no other explanation seemed to work.

Captured by x-ray telescopy, Cygnus X-1—the blue mass in the picture at right—emits plentiful x-rays but no light and has the mass of ten suns. This stellar body may be a black hole, a kind of collapsed star in which, according to relativity theory, time and space are obliterated.

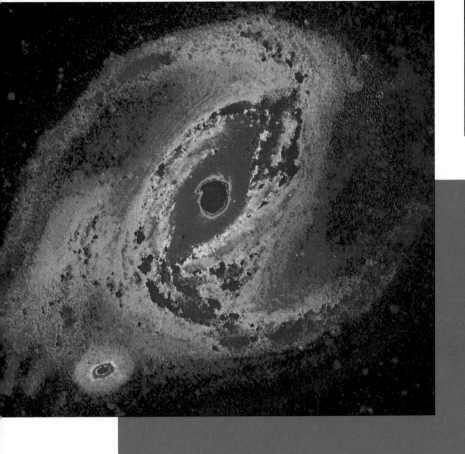

sure that his intuitive sense was correct, and he was able to back it up with mathematical proof. In 1905, shortly before his famous article on special relativity appeared, he published a compelling explanation of the photoelectric effect inspired by Planck's obscure theory.

In 1916, American physicist Robert Millikan substantiated Einstein's discovery, and the light-as-particle theory finally began to win grudging acceptance. Einstein's light particles were given the name photons, and they came to play an essential role in the developing theory of quantum mechanics.

The rules of quantum mechanics were hammered out during the early part of the twentieth century by physicists working in many lands. Foremost among them was Danish scientist Niels Bohr, whose laboratory in Copenhagen became a hub of atomic research. Bohr proposed that the behavior of light quanta is dictated by the structure of the atom itself. Negatively charged particles in an atom, he

In fact, Einstein seems to have been one of the very first persons to take Planck's idea seriously. In 1904, while he was wrestling with his theory of special relativity, Einstein was also trying to make sense of the mysterious photoelectric effect—a puzzling phenomenon in which

Spiraling red jets emanate from the dark core of galaxy NGC-1097 in this computer-colored image and could come from a supermassive black hole spinning rapidly enough to sling matter from its center out along the axis of spin. Astronomers speculate that there may be black holes at the heart of many galaxies, including our own.

metal sheets exposed to light give off charged particles. Working with Planck's idea, Einstein came to realize that if he extended the quantum concept to embrace light itself— that is, if light consisted of particles instead of waves—the photoelectric effect would be perfectly logical. It was considered an outrageous idea at the time, but Einstein was

suggested, do not whirl around the atom's nucleus in continuous orbits as previously thought. Rather, they are confined to specific positions, or "stages," relative to the nucleus. Their only means of releasing their quanta of energy, in the form of light or other radiation, is by jumping from one stage to a lower one. This notion gave origin to the term *quantum leap,* which is now a staple of everyday language.

Probing further, and armed with ever more powerful experimental tools such as the newly invented particle accelerators, physicists have contrived to dissect the atom. The classical model of a solid nucleus surrounded by orbiting electrons was replaced by an infinitely more complicated one. New and smaller particles were discovered so frequently that physicists found they needed to refer to a booklet called *Particle Properties Data Handbook* simply to keep track. And the discoveries have continued to this day.

The current subatomic bestiary lists particles with such exotic names as gluons and muons, gravitinos, bosons, and neutrinos, down to the very smallest of all, the enigmatic leptons and quarks. It is possible that there are even smaller fragments within the atom, but finding them becomes increasingly unlikely. One writer put the difficulties in perspective by making this analogy: If an atom were the size of the earth, any component part of a quark or lepton would have to be smaller than a grapefruit to have escaped detection.

Of all the elemental particles, the elusive bits known to scientists as quarks have the most topsy-turvy charm. Individual quarks have never been isolated; they apparently travel in bunches of three. Physicists have divvied these particles, which they believe to be the building blocks of the atom's nucleus, into six categories, or "flavors," with the whimsical names "up," "down," "strange," "charmed," "bottom," and "top." Each flavor comes in three supposed "colors"—red, green, and blue. Even the name "quark" has an offbeat background. It was borrowed from the cryptic phrase "three quarks for Muster Mark," in James Joyce's *Finnegans Wake.*

Even though atomic bits smaller than quarks and leptons may never be found, some quantum theorists question the assumption that the known particles are really the end of the line. American physicist John Wheeler speculates that underlying all of space-time is a sea of quantum foam, a nebulous energy-matter mix made up of "all the time and everywhere." Perhaps, Wheeler suggests, charged electrons are not really quantum fragments, as we think of them now, but are the ends of tubes connecting our universe to everything that underlies it.

 ong before quantum physicists divided their world into so many components, they faced a dilemma that seemed insoluble and led to a new scientific doctrine called the uncertainty principle. To "observe" and measure any elemental particle, the physicist has to bounce it off some other quantum of radiation and observe the results. An inherent problem in such a process is that the collision inevitably alters the particle under study, by knocking it off course, slowing it down, or somehow changing its shape. Thus, any single experiment can determine either a particle's momentum or its position but never both simultaneously. Defining the one variable permanently distorts the other.

Physicists came to realize early on that this difficulty cannot be resolved by improving the measuring techniques. They recognized that the elusive behavior of subatomic particles must be treated as a permanent fact of life in the study of the quantum universe. Physicist Werner Heisenberg, who worked for a time with Bohr in Copenhagen, spelled out the uncertainty principle in the 1920s, and his tenet still lies at the heart of quantum mechanics. He pointed out that a precise description of the movements of individual subatomic particles is quite simply beyond our reach. Although predictions of particle behavior based on statistical probability can be remarkably accurate, we can never know with absolute certainty what an individual electron, proton, or quark will do or when it will do it.

In fact, the uncertainty goes even deeper. Quantum laws point out that an electron or any other particle in its natural state has no specific identity or agenda. It exists in a netherworld of potential being—as part of what is called a wave function. Only when an external event, such as the physicist's experiment, intrudes on this condition does the wave "collapse." And only then does the electron firm up into one of its possible states of size, energy, or time.

What Preceded the Big Bang

As scientists study the origins of the universe, the answers they reach often lead to fresh questions. Assuming that time, space, and the present cosmos burst into being with the big bang, the inevitable question arises: If the big bang brought the universe into existence, what existed before that? Exactly how did nothing become something? While some say the problem belongs in divine hands, others persist in their calculations.

Some physicists say there was no "before," because the universe expands and contracts in an infinite cycle, without end or beginning. But many believe the universe is running down, and all will at last come to rest. If so, the number of cycles before this one is finite—and the original question remains. Quantum mechanics, the study of subatomic particles, has led some theorists to posit that the universe did arise out of nothing—although their definition of "nothing" is very different from ours.

To quantum physicists, "nothing" is rich and fertile stuff: Even in "empty" spaces, at subatomic levels and in infinitesimal bits of time, elementary particles pop in and out of existence, in what are called vacuum fluctuations. "Empty space," notes physicist John Wheeler, "is not empty. It is the seat of the most violent physics."

Given such busy "emptiness," physicist Alexander Vilenkin posits a primeval—and continuing—"space-time foam," in which tiny submicroscopic universes continually begin to exist, and cease, in a restless stew of vacuum fluctuations. Most of these bubble universes fizzle out, but given the laws of probability, once in a while, one is viable. In fact, ours may not be the only bubble that has become a universe. Others may exist alongside ours, but in another space-time, so that they cannot be detected.

Astonishingly, a vacuum fluctuation violates no laws of physics. MIT Physicist Alan Guth describes the first expansion of the primeval, proton-size universe as a phase change—like the change of hot water into steam—in which energy became matter. Guth says it was "the ultimate free lunch."

These rich theories of nothingness seem more akin to science fiction than science fact. Wheeler even speculates that one day humans may have the means to create a universe. "After all," he says, "for everything we know, our universe could have been made in somebody else's basement."

Both Max Planck (top) and Niels Bohr broke away from classical physics to make brilliant discoveries in the behavior of subatomic particles. Einstein compared Bohr's insight to a miracle, calling it "the highest form of musicality in thought."

Although physicist Erwin Schrödinger was one of the chief figures, along with Heisenberg and Bohr, in the development of quantum physics, he found the implications of the uncertainty principle extremely difficult to swallow. He illustrated his misgivings by inventing a paradox that has become known as Schrödinger's cat. In this scenario, an imaginary cat is put into a box along with a device that might or might not let off a deadly poison, depending on whether or not a single radioactive atom begins to decay. If the atom decays, the poison will be released and the cat will die; if the atom does not decay, the cat will live. The experiment begins when the box is closed and left unattended.

Quantum uncertainty holds that until an atom is observed, it exists in a hybrid situation of all potential states. This means the radioactive atom in Schrödinger's fictitious box will exist simultaneously in states of decay and nondecay until someone investigates the results of the experiment. By implication, therefore, the paradox suggests that, in violation of all common sense, the cat must be both alive and dead until the box is opened.

The paradox is a good one, and the ambiguities it illustrates led many scientists to question the validity of quantum mechanics. Einstein, convinced that every action has a cause, was one of the chief objectors. He voiced his distrust of quantum's uncertainty in a letter to his friend Max Born. "I find the idea quite intolerable," he wrote, "that an electron exposed to radiation should choose of its own free will, not only its moment to jump off, but also its direction. In that case I would rather be a cobbler, or even an employee in a gaming-house, than a physicist." He held firm to this deterministic position.

Niels Bohr had no prescription for Einstein's discomfort with the inherent uncertainty predicated by quantum theory. Since the theory successfully predicted the outcome of atomic behavior, he questioned the need to know the details of every atomic event. "It is wrong," Bohr once said, "to think that the task of physics is to find out what nature is. Physics concerns what we can say about nature." Still, he did not belittle the theory's perplexing difficulties. "If anybody says he can think about quantum problems *without* getting giddy," he wrote, "that only shows he has not understood the first thing about them."

Bohr seemed to be suggesting that in science familiarity breeds humility. And there is no doubt that the centuries of scientific inquiry have led humankind to a vastly diminished view of its role and importance within the universe. Each new discovery about space and time has moved us further from our self-proclaimed place at the center of the universe. In the modern scientific view, we are minuscule organisms clinging to a tiny speck in the vastness of the cosmos.

Such realism notwithstanding, nothing seems to temper the timeless dream of exploring the cosmos. For the moment, it seems, our technology cannot even begin to keep pace with this dream because the distances of space are simply too great. Traveling at 61,148 MPH, *Voyager 2*

took twelve years to reach the planet Neptune, within our own solar system. At that rate, *Voyager* would need about 50,000 years to reach Proxima Centauri, the closest star in our galaxy. And trips to other galaxies are more difficult to ponder. Even if *Voyager* could reach light speed, it would take two million years to get to neighboring Andromeda.

Within the known limits, however, some intriguing ef-

Tracks of charged subatomic particles, recorded in a bubble chamber and later color enhanced, may hold clues to the events of the big bang and to a grand unified theory (page 63) that would account for all the forces in the universe.

fects still seem possible. If humans learn to travel at only half the speed of light, this will entail important changes in our normal sense of order and time. Time does not stand still for space travelers, but relativity shows that it does slow down—thus affecting the aging process. An oft-cited example of this theoretical impact involves the disproportionate aging of a pair of twins, one of whom stays on earth while the other travels through space. Moving at 98 percent of the velocity of light, the twin in space would experience just ten years while fifty-one years passed back on earth. When the space twin returned, he would be forty-one years younger than his brother, although neither of them would have noticed any change in the normal passage of time. The longer or more frequent the journeys in space, the more dramatic the discrepancies in time. A veteran space traveler, at the end of his career, might retire on earth the same age as his earthbound great-grandchildren.

The one shortcoming in this apparent prospect for a technological Fountain of Youth is that although space travel can slow time down, it cannot reverse it. Space travelers may stay in their twenties for ages compared with the folks back home, but no matter how fast they travel, they cannot return to earth any younger than when they departed. Even this boundary, however, does not seem to be absolute.

Scientists have begun to explore possible shortcuts through space-time that would enable us to travel instantaneously from one part of our universe to another. Such theoretical tunnels in and out of the cosmos have been called wormholes, in reference to the passageways created by a worm eating its way through an apple.

Wormholes have never been proved to exist, but the possibilty has won the endorsement of some highly respected scientists. Alan Guth of the Massachusetts Institute of Technology argues that the idea is theoretically possible. Relativity dictates that massive objects distort the space-time around them. Guth and others suggest that the distortion may take the form of something like a tube, providing access to a distant place in the universe and to another framework in time. Stephen Hawking has proposed that if wormholes exist at all, they may be present in countless numbers, scattered throughout the universe. Another theoretical possibility for transversing time involves tachyons, hypothetical particles that, if they actually exist, move faster than the speed of light and therefore backward in time. Gregory Benford, now at the University of California, Irvine, collaborated with two colleagues in a 1970 issue of *Physical Review* to describe a "tachyonic antitelephone." They suggested that if a beam of tachyons could be somehow modulated to carry a signal, it could transmit a message to the past. Of course, it could work only if those to whom the transmission is aimed have a tachyonic receiver.

Scientists are thus undeterred in their search for detours to the limits of space-time. The theoretical possibility of shortcuts within our universe from one time frame to another would seem to open up many other possibilities. There would be nothing, for example, to rule out the existence of comparable passageways to other universes. Within some circles of big bang theorists, it has long been acknowledged that such alien universes might very well exist. And—as if matters were not already complicated enough—some scientists have come to believe that our four dimensions may be just the detectable few among many. Struggling alongside Stephen Hawking in his effort to unite the theories of relativity and quantum mechanics are physicists who study theoretical entities called strings. This term is used to describe the behavior of particles under high-energy conditions, such as those that are thought to have prevailed at the start of the big bang. String theorists are convinced that there must be either ten or twenty-six dimensions, most of them somehow hidden from our awareness.

No one really believes that even the combined efforts of cosmologists, physicists, astronomers, and mathematicians will resolve anytime soon all the mysteries of space and time. On the contrary: In science, the exciting thing is that every new discovery seems to unfold new mysteries. "The most beautiful experience we can have," said Albert Einstein, "is the mysterious. It is the fundamental emotion that stands at the cradle of true art and true science."

Ways to Travel through Time

Humans have long dreamed of being able to travel through time—to find ways to remain eternally young, correct misdeeds of long ago, glimpse what fate holds in store. Yet all but the most nimble minds have considered movement through time a fantasy. Indeed, it is the stuff of countless science-fiction novels. Over the last few decades, however, in a case of truth being nearly as strange as fiction, a number of scientists have explored just how time travel might come to pass.

Time travel has always made scientists uneasy because it allows something to affect its own past. That violates causality, the basic principle of cause and effect. Consider, for example, the often-cited grandfather paradox: If you could travel back in time, you could kill your grandfather. But if you killed your grandfather, you would never be born. And if you were never born, you could not have traveled back in time to kill your grandfather. So, who killed your grandfather? Violating causality breaks the laws of science, and that is all many scientists need to discount time-travel theories. But the three "time machines" described on the following pages cannot be so quickly dismissed. They are based on Einstein's theories of general and special relativity, and although each requires the use of technology that goes far beyond our dreams, they represent possible, if difficult and far distant, means of crossing humankind's last unchallenged frontier. They also represent, to mathematicians and physicists, the limits and the possibilities of science as we know it.

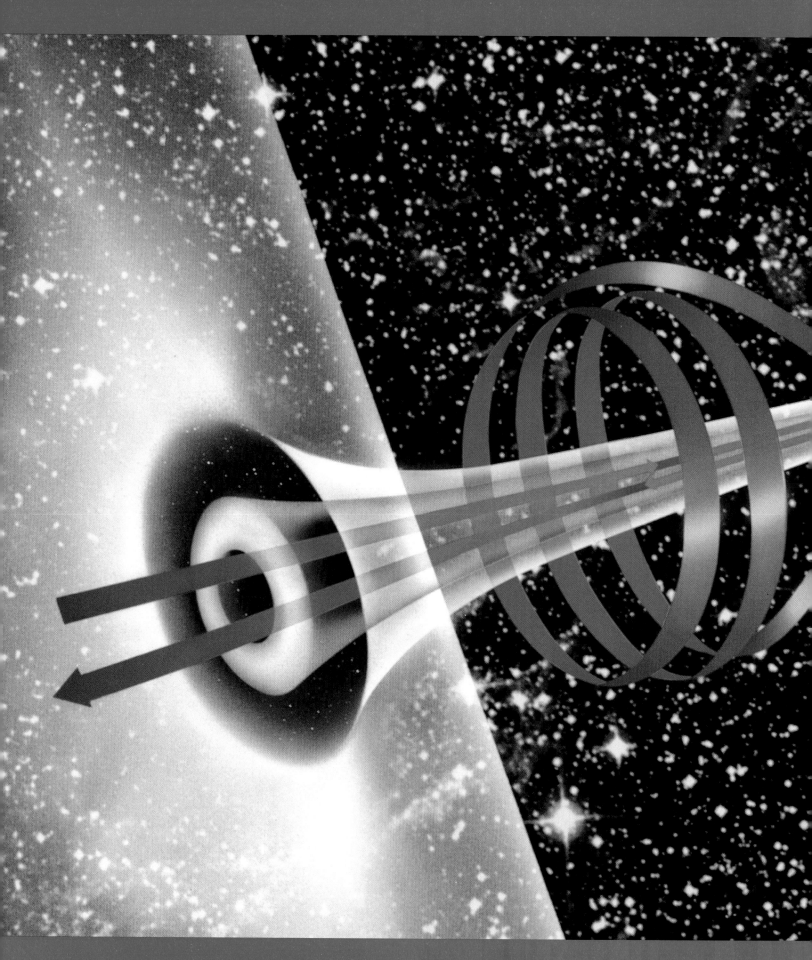

Journey through a Spinning Singularity

According to Einstein's theory of general relativity, if enough mass is collected in one place, its gravity field will warp time and space as we know it. In this cosmic wrinkle, some scientists reason, time travel would be possible. Any object producing such a mega-gravity field would be like a time machine, and some experts speculate that a black hole is such an object.

Black holes are thought to form when a star burns out and collapses on itself. This creates a singularity, a point of infinite density and gravity so strong that even light could not escape. Surrounding the singularity is the event horizon, a no man's land in which the hole's gravity could be felt and beyond which nothing could be seen because all light would have been absorbed.

Physicists have used Einstein's field equations to develop several black-hole models. Most of them would not be navigable—any object approaching the event horizon would be sucked into the hole and instantly crushed to infinite density. The only black hole deemed theoretically navigable is also the one thought most likely to form in space—the rotating black hole.

The concept of the rotating black hole was put forth in 1963 by mathematical physicist Roy Kerr of the University of Texas. All stars spin and when a star collapses, the spin influences the resulting hole's structure; there may be a number of event horizons, and the collapsed star becomes oblate, much like a dinner plate. If the spinning hole could be formed into a ring and kept stable, says Kerr, this formation would, in theory, allow time travel. A spaceship could cross the event horizons and steer through the ring singularity *(right)*, emerging in hyperspace, where time-space coordinates have been warped by the hole's spin. Moving through space is now comparable to moving through time.

In hyperspace, the pilot navigates a circular path near the spinning singularity. Traveling with the hole's rotation sends the ship into the future; moving against it sends the ship into the past. Once the desired time period is reached, the pilot leaves hyperspace and zips back through the ring to our space—in the chosen time *(above)*.

Outer Event Horizon

Ring Singularity

Inner Event Horizon

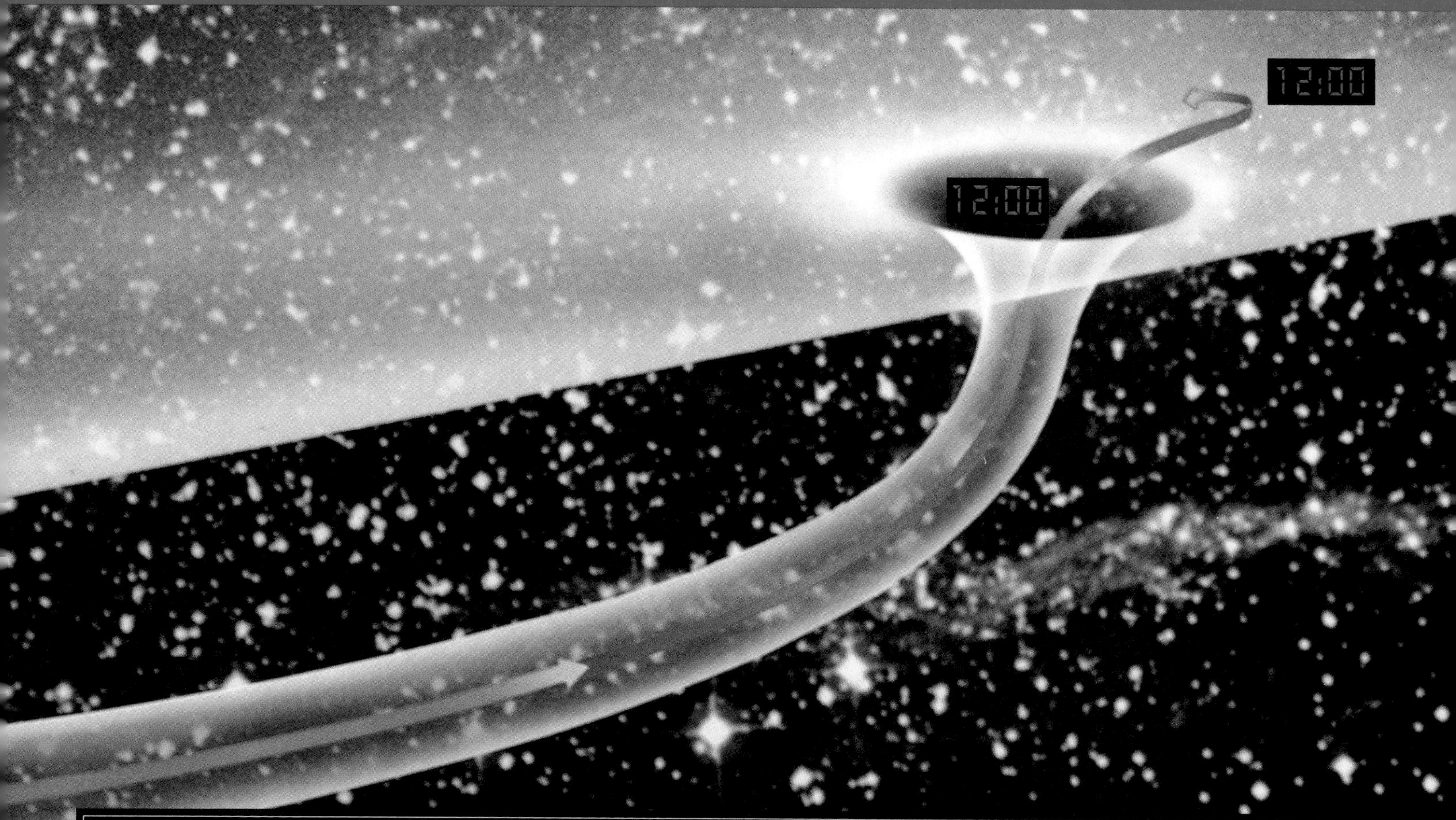

Tunnels into the Past

Cosmic wormholes exist only in science fiction and mathematical equations. But this fantasy freeway to new time and space dimensions has gained attention in recent years because of the work of two scientists who used those equations and Einstein's theories to develop a time-travel scenario.

The cosmic wormhole is named for the shortcut a worm takes by boring straight through an apple rather than crawling across it. Similarly, a cosmic wormhole tunnels through space-time and leads from one region to another. According to California Institute of Technology physicist Kip S. Thorne and his former assistant, Michael S.

Morris, now at the University of Wisconsin, the first step toward wormhole time travel is to find an existing wormhole, then use some form of "exotic matter" to prop open its mouths. From here, special relativity takes over.

According to this theory, the closer an object comes to light speed, the more time slows for that object. Thus, an astronaut who left an identical twin on earth and traveled into space at close to light speed would return to find that he or she had aged at a slower rate than the stay-at-home twin.

Think of the wormhole mouths as the twins. If one mouth can somehow be accelerated to near light speed—

moving in a tight circle, as seen in the diagram at lower left—and then stopped, it is comparable to the traveling, or "younger," twin. A clock inside this mouth will have advanced at a slower rate than one at rest just outside the mouth. However, clocks resting inside the stationary mouth and just outside it would also show the slower rate. In other words, the two wormhole mouths open onto parts of the universe that exist in different times. So, if a spaceship flies into the mouth that moved, it travels back in time, emerging from the stationary mouth perhaps in time to see itself entering the mouth that moved *(above)*.

Alas, the cosmic wormhole is not the ultimate time machine. Thorne and Morris say it allows visits only to the past, and then only as far back as the wormhole's creation. And neither physicist knows how a wormhole is formed. But they point out that nothing in the laws of physics prohibits one

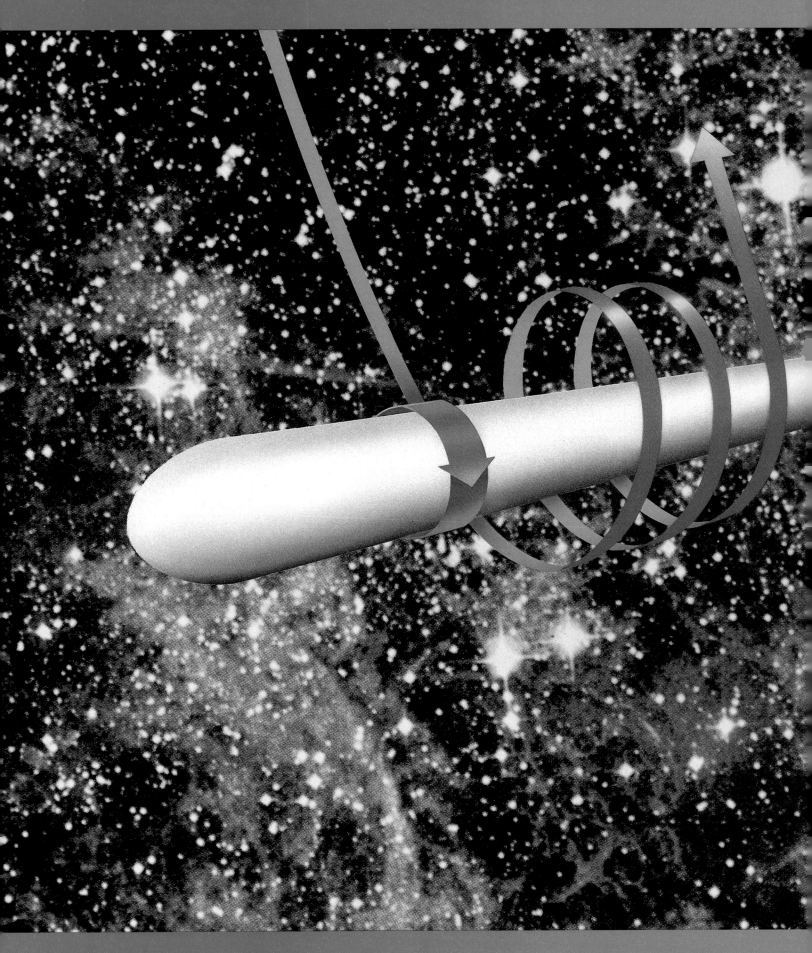

A Giant Cylinder That Warps Space-Time

If scientists contemplate time travel at all, it is usually not because they believe it may be possible someday but because of what can be learned about the laws of physics by constructing a theory. However, there are people who believe great leaps forward and backward through time will ultimately be possible, especially with the aid of special devices that would be constructed in the far distant future. The prospect that it may take another millennium before the technology exists to build those devices does not deter such forward thinkers as Frank Tipler, a Tulane University mathematician.

Tipler's design found its genesis in Einstein's theory of general relativity, which states that sufficient mass can change or distort space-time through gravity. Extending Einstein's ideas a little further, Tipler theorized that if a massive object in space rotates at a velocity of half the speed of light, space-time could be twisted

upon itself, much like a milk shake in a soda-fountain blender. This means that two points that were once far apart in time are now brought close together—in fact, they are in the very same place. That is the point at which time travel becomes possible.

According to Tipler, time travel can be achieved, at least mathematically, if a future generation creates a massive cylinder whose length is approximately ten times greater than its diameter. Rotating the cylinder at 93,000 miles per second, or half the speed of light, would create a time-warped area of space around the center of the cylinder. What would-be time travelers have created, then, is essentially a man-made rotating black hole—the spinning cylinder acts much like the hole's rotating ring singularity.

To use the cylinder as a time machine, a spaceship must simply maneuver to the cylinder's midpoint and navigate around it—moving against its

rotation to travel into the past and with its rotation to journey into the future. Each revolution would bring the spaceship further back or forward in time. When the time travelers reached their destination, they would merely have to speed their ship away from the cylinder. As with all theoretical time machines, however, there is a caveat. Like the wormhole, you cannot go any further into the past than the time of the cylinder's creation.

Although Tipler's cylinder is theoretically possible, he and other open-minded scientists admit that "we are a very long way from completely resolving the causality violations question" raised by time travel. Yet as astrophysicist and writer John Gribbin noted in his 1979 book, *Timewarps,* if rational minds summarily dismiss such things as time machines on the basis of acausality, that "tells us more about the human mind than about the physical nature of the Universe."

CHAPTER 3

The Human Perspective

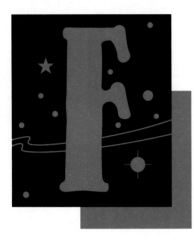

or four months, from January 13 to May 22, 1989, Stefania Follini existed outside of time. The twenty-seven-year-old Italian interior designer did not see the sun rise and set, feel the daily temperature rise and fall, or witness the change of seasons. She consulted neither calendar nor clock. Follini was living deep within the earth, in a cave near Carlsbad, New Mexico, making her home in a twenty-by-twelve-foot Plexiglas module. Inside, the temperature registered a constant sixty-nine degrees, and the artificial light could be dimmed but not extinguished. Alone, a computer her only means of communicating with the scientists monitoring her, Follini was a human guinea pig in an experiment designed to study the mental and physical effects of living in an environment completely cut off from all time cues.

Without the means to measure hours or days, Follini lost track of time. Instead of following a twenty-four-hour cycle of sleeping and waking, she increasingly stretched out her "day" to twenty-five hours, then twenty-eight, and eventually to about forty-four hours. Sometimes she would be active for as many as thirty hours; at other times she would sleep for up to twenty-four. As a result, when the experiment was over and she finally emerged from her subterranean home into the bright New Mexican sunlight, Follini believed that she had been underground only two months. In actuality, 130 days had passed.

Similar experiments with other volunteers have shown that most people isolated from time cues experience a less extreme lengthening of their day—to about twenty-five hours. But all conclude that without the means to measure time, humans tend to "free-run," shaping their days according to their own inner schedules. And the sleep-wake cycle is not the only bodily function that keeps to the beat of an inner rhythm: Rate of respiration, rise and fall of body temperature and blood pressure, and the frequency of cell division are all governed by biological clocks.

Yet even as the body marches along to nature's steady beat, an individual's psychological perception of time may register an entirely different tempo, fluctuating according to other factors, becoming distorted by events that occur every day. When a person is elated, or concentrating intensely, for example, time seems to fly; but when the same individual is idle, or performing a dull or repetitive task, time drags. And our sense of time's passing changes as we age. Days may seem almost boundless for a child, but they pass all too quickly for an adult.

Internal rhythms and psychological perceptions are not alone in shaping our sense of time; it is likewise affected by social cues. Cultural influences such as mythology, religious beliefs, philosophy, and scientific principles also help form a person's view of time. Different cultures and societies have different ways of thinking about time. In some societies, time is considered as flowing along a linear path, with one event following another, stretching infinitely into the future. The pace of life is frequently hectic; each day is scheduled, filled with goals to be accomplished in a predetermined amount of time. Time is equated with money, and idle moments are wasted resources. Other societies tend to reflect upon the past and tradition when considering what is to come, rather than rushing headlong toward the future and progress. They are more relaxed about the amount of time it takes to accomplish a given task.

Ultimately, these and other factors combine to influence our perception of time. But perhaps the most remarkable among them is the human body, which is in many ways an extraordinary timekeeper. Body temperature rises and falls a degree or two each day with clocklike precision. A woman's menstrual cycle returns month after month. The body's supply of HGH (Human Growth Hormone) climbs to an annual peak each summer, which explains why children grow fastest during that time of year.

These and dozens of other biological rhythms—or biological clocks, as they are sometimes called—are now being closely studied under a relatively new scientific discipline known as chronobiology. Scientists are finding that every biological function within the body, from the split-second pulsing of brain waves to the annual rise and fall of serotonin, the hormone that regulates moods and sleep, follows distinct cyclic beats. Nor are these rhythms unique to humans. All living things, including plants and single-cell organisms, possess some kind of internal timekeeper.

Biological rhythms are incredibly complex. As scientists studying Stefania Follini and other volunteers like her found, the rhythms not only work in tandem with one another, they also keep step with the cycles of the natural world—the alternating periods of daylight and darkness, the waxing and waning of the moon, the changing seasons. In fact, chronobiologists believe that biological rhythms originally evolved as a direct response to the light and dark cycle of the sun. Early humans and other creatures needed to be able to anticipate daily and seasonal changes in sunlight. The nightly decline in adrenaline and other "wakening" hormones, for example, ensured that our ancestors would be quiet at night, when they were at greatest risk of harm from predators. And the slowing of the body's metabolism in the fall ensured that those same ancestors would have extra fat on their bodies during the cold and food-scarce winter. Scientists also speculate that the internal clocks of birds and other migratory animals remind them when to set off in search of winter feeding grounds and when to return in spring to breed (page 82).

Biological rhythms

Although de Mairan's observation was correct, his conclusion was incorrect. The plants inside of the shed could not have sensed the sun without seeing it; no plant is that sensitive. But the plants could, even when isolated in darkness, keep in step with the sun's daily cycle—strong evidence of the inherited nature of biological rhythms.

De Mairan's experiment was largely ignored until recently. During the past few decades, as chronobiology has grown into a full-blown science, more and more researchers have followed up on his work, often with exciting and unexpected results. They have discovered, for example, that few biological rhythms—in plants or animals—exhibit a daily cycle that coincides exactly with the twenty-four-hour cycle of the sun. For this reason, chronobiologists use the term *circadian,* from the Latin for "about a day," to describe daily rhythms. Interestingly, animals that are active during the day (including humans) tend to have circadian cycles that run slightly longer than twenty-four hours; animals that are active at night, on the other hand, usually have cycles that run slightly shorter than the solar day. The explanation for this difference still eludes scientists.

In addition, chronobiologists have discovered that seven-day cycles exist throughout nature, from bacteria and other simple forms of life to all species of mammals, includ-

may have initially required a direct dose of natural light to help them maintain a steady pace. They have since evolved, however, to the point where they can hold their natural tempos without the influence of the sun. This was first illustrated by the French astronomer Jean Jacques d'Ortous de Mairan. In 1729, he conducted a small experiment with a species of mimosa known commonly as the sensitive plant. Its leaves normally open during the day and close at night. Wondering what the plant would do if placed in constant darkness, de Mairan shut several mimosas up in a dark shed. Later he reported his findings to France's Royal Academy of Science. "It is not at all necessary," he wrote, "that it be in the sun or open air. The phenomenon is merely a little less marked when the plant is kept shut up in a dark place; it still opens out very noticeably during the day and folds up or closes regularly in the evening, for the entire night. The sensitive plant thus senses the sun without seeing it in any way."

ing human beings. For reasons that remain unclear, blood pressure, heartbeat, and body temperature have all been shown to have weekly as well as daily rhythms. The rise and fall of several body chemicals, including cortisol, the hormone that helps the body combat stress, also have weekly cycles. Most astonishing has been the discovery that the body's immune system seems to become vulnerable in seven-day intervals. Long before penicillin came into use, doctors were acutely aware that pneumonia and malaria patients were at greatest risk around the seventh day of their illness. Today, surgeons who perform organ transplants know that the immune systems of their patients are most likely to reject the new organs in seven-day cycles.

While daily and seasonal cycles have an obvious link to the sun, and monthly cycles appear connected to the moon, nothing in nature is clearly responsible for a weekly rhythm. Some scientists speculate, however, that our seven-day social week may be rooted in a biological source.

Whether a particular cycle lasts one day or seven, humans must try to compress the body's slightly longer biological day into a twenty-four-hour schedule. Fortunately, the body is able to reset its daily rhythms with the aid of powerful time cues, which chronobiologists call *zeitgebers* (German for "time givers"). The sun is by far the most important zeitgeber. Each morning the sun's appearance in the eastern sky cues the body's rhythms to reset and begin their daily cycles—even though they would have preferred to wait another hour or so to kick in.

The body's rhythms take cues from other zeitgebers as well, such as the daily and seasonal rise and fall of air temperature and the monthly changes in the moon's gravitational pull. Some scientists believe electromagnetic fields may also help synchronize our internal rhythms. But not all zeitgebers are found in nature. Habits such as going to bed at the same time each night, rising for work at a set hour each morning, or eating meals at the same intervals each day can serve as cues. In addition, the internal rhythms help synchronize one another, working together throughout the day like members of a well-rehearsed orchestra.

Under experimental conditions, people can sometimes reset their rhythms to an arbitrary "day" that is markedly shorter or longer than the twenty-four-hour day established by the sun. In 1938, for example, sleep expert Nathaniel Kleitman and his associate, Bruce Richardson, spent thirty-two days in Mammoth Cave, Kentucky. The two scientists attempted to put themselves on a twenty-eight-hour day, staying awake for nineteen hours and then sleeping for nine. Richardson, who was twenty-three years old at the time, adjusted to the longer day within a week. But Kleitman, who was forty-three years old, experienced difficulty with the new schedule; he was frequently sleepy and irritable during the day and slept fitfully at night. Records of their daily body temperature cycles revealed that Richardson's temperature regularly peaked during his waking hours in the cave and dropped while he slept; he had successfully reset his biological clocks to the twenty-eight-hour day. Kleitman's temperature cycle, in contrast, did not adjust but clung stubbornly to a roughly twenty-four-hour rhythm.

More recently, a group of volunteers and scientists spent several weeks on the Spitsbergen Islands above the Arctic Circle. They went during mid-summer, when the sun never sets and thus when the natural rhythm of night and day is absent. The volunteers divided into two groups. One group set its mechanical clocks to a twenty-one-hour day, the other to a twenty-seven-hour cycle. Both groups soon adjusted to their new schedules, as evidenced by changes in body temperature and other circadian rhythms.

cientists can only speculate on why some people adjust more easily than others to a shorter or longer day. Curiously, personality type seems to be one factor. Extroverts tend to have more flexible rhythms than introverts. This may be linked to their body temperature cycles. The cycles of extroverts are generally more variable, making it easier to adapt to a new schedule. Age also seems to influence the flexibility of a person's rhythms, as Nathaniel Kleitman discovered; the older a person is, the more rigid his or her rhythms seem to be.

The Great Migration Mystery

As winter approaches and again as it recedes, hundreds of millions of birds throng the skies along the world's great migration routes. But birds represent only a fraction of the animal kingdom's migrants. Throughout the year, creatures as diverse as butterflies, zebras, and sea turtles demonstrate regular patterns of long-distance travel.

What triggers their seasonal treks? Scientists have addressed the question, and the results reveal that an innate, precise sense of time and space may not only help initiate the animals' migration but also help them navigate the journey.

The signals that control bird migrations are still a mystery. For some species, weather-related conditions, such as day length, are thought to initiate a biological reaction that causes development of sex organs, the deposit of fat to fuel long flights, and restlessness. Yet in other species, internal timekeepers produce the same response without outside stimulus.

Birds may also have an innate sense of direction. Early students of avian navigation believed that birds, like coastal sailors, relied solely upon landmarks for guidance. But studies in the 1940s and 1950s showed that at least some birds have an internal compass.

In one experiment, German researcher Gustav Kramer put starlings in round cages and watched them become restless as time for migration arrived. With their view blocked except for the sun, the caged birds faced in the proper direction for the trip they could not take. When Kramer used mirrors to change the direction of the sunlight, the birds adjusted their position. The starlings not only take bearings from the sun, Kramer found in further experiments, but also rely on remarkably precise internal clocks, which allow them to compensate for the movement of the sun.

Kramer's findings revolutionized ornithology and led to additional discoveries. Some birds, scientists found, use the stars as well as the sun to determine direction. Furthermore, the earth's magnetic field seems to play a role in avian navigation; other factors may be qualities of light and sound beyond the range of human senses.

But if scientists are beginning to understand how animals navigate, they still have no definite explanation, in many cases, for why the creatures go where they do. There is no clear biological reason, for example, why some sea turtles return yearly to specific beaches to lay their eggs.

And no one can say why some migratory journeys are so much longer than seems necessary. Arctic terns depart from their polar haunts with the onset of each winter. But these birds do more than escape the weather; they fly to the Antarctic, passing through temperate zones and tropics en route. After only two months' respite, the birds begin the return trip so they will reach their northern habitat in time to breed. Every year arctic terns, driven—and guided—by their internal clocks, travel for eight months and cover some 25,000 miles.

Lesser snow geese head for their spring breeding grounds in the Arctic. The long trip north consumes layers of fat, built up during months of feeding and, perhaps in part, as a result of the rhythmic cycle that prepares birds for migration.

North American monarch butterflies (below) cluster on a tree in Sierra Chinqua, Mexico, at the end of their autumn migration. Some of the colorful insects travel more than 1,200 miles to their winter quarters. So predictable is their time of migration that towns along the butterflies' route plan festivals to celebrate their arrival.

Pacific turtles lumber ashore at Ostional Beach, Costa Rica. At the height of the annual invasion, tens of thousands of them cross the beach, laying and burying their eggs and returning to the sea before daybreak. Scientists speculate that the moon or the turtles' biological clocks may trigger the landing.

Most people do not become fully aware of their biological rhythms until their cycles are thrown out of sync. Traveling rapidly across several time zones or switching to a different work shift are two of the most notorious ways of disrupting one's internal time clock. In each case, the body must adapt to a new set of time cues. Recovery comes only gradually, as one by one the body's rhythms realign themselves; the changeover can take many days to complete. The rhythms that affect alertness and concentration may take two days to two weeks to resynchronize. Heart rate, which typically slows by a few beats at night, often needs five days to adjust to a new circadian cycle. And urine output, which is usually lowest at night no matter how much liquid is consumed during the day, can take up to ten days to shift into a new schedule—the reason long-distance travelers often find themselves waking up in the middle of the night to go to the bathroom.

Jet lag—the rhythm disruption that comes with the crossing of several time zones—is, of course, a relatively new phenomenon. Back in the days when oceans were crossed by tall ships rather than by supersonic jets, travelers passed from one time zone to another over a period of many days, enabling their bodies to adjust slowly to new time cues from the sun. Today, people jet across several time zones in a matter of hours, a disrupting experience for the body's rhythms. Studies have shown that the human body cannot adjust to changing time zones much faster than two hours a day. So even a flight from Los Angeles to New York—a crossing of three time zones and a time change of as many hours—can leave travelers feeling tense, tired, and slightly out-of-sorts for a few days as their rhythms reset themselves to New York time.

Before the invention of mechanical clocks, when navigators told time only with the stars and the sun, sailing crews traveling long distances were often unaware that their biological clocks were resetting themselves. Those who circumnavigated the globe, however, would learn upon returning home that they had either "lost" or "gained" an entire day. Such was the case with Portuguese explorer Ferdinand Magellan's famous three-year voyage around the world during the early 1500s. When the *Victoria*, the

Children in Murmansk receive a dose of artificial sunlight, a routine measure in the northern Soviet Union. The treatments are deemed essential to stave off vitamin deficiencies during the nearly fifty days of polar night. Another reason for controlling indoor lighting in the region is to prevent disturbance to the sleep cycle, normally regulated by the sun.

last remaining ship of the explorer's ill-fated fleet, arrived back in Seville, Spain, on September 8, 1522, its log recorded the date as September 7. A day had vanished from the ship's calendar.

The *Victoria*'s crew had first become aware of the lost day two months earlier when they had briefly anchored off the Canary Islands. "In order to see whether we had kept an exact account of the days, we charged those who went ashore to ask what day of the week it was," wrote Antonio Pigafetta, the chronicler chosen by Magellan to record the events of the voyage. "They were told by the Portuguese inhabitants of the island that it was Thursday, which was a great cause of wondering to us, since with us it was only Wednesday. We could not persuade ourselves that we were mistaken; and I was more surprised than the others, since having always been in good health, I had every day, without intermission, written down the day that was current."

The *Victoria* had been traveling westward, or with the sun, which meant that each day ran slightly longer than twenty-four hours. Pigafetta and the other crew members, of course, had no electronic timepiece to tell them their days were running long. Only when the sailors reached home did they accept that their trip had cost them a day, a quirk of time that would eventually prove so annoying to global commerce that the International Date Line was established in 1884 so the world could agree on where each calendar day begins *(page 110).*

The experience of the missing day may have perplexed Pigafetta's mind, but it did not affect his body. He had lost the day gradually, without any serious disruption to his internal rhythms. In fact, no one really noticed the physiological effects of moving across time zones until after the airplane was invented approximately 400 years later. The first aviator to experience jet lag was Wiley Post, an American adventurer who in 1931 flew his single-engine propeller airplane, the *Winnie Mae,* around the world in eight days. Before making the historic flight, Wiley had predicted that flying through twenty-four time zones in a little more than a week would "bring on acute fatigue if I were used to regular hours. So, for the greater part of the winter before the flight, I never slept during the same hours on any two days in the same week. Breaking oneself of such common habits as regular sleeping hours is far more difficult than flying an airplane."

By mixing up his sleep-wake cycle, Post may have accustomed his biological rhythms to a more flexible schedule. Few travelers, however, would care to follow such a rigorous training program. Today's jet-lag experts recommend a modified approach, suggesting that passengers begin to shift their sleep cycle a day or two before an overseas flight. Passengers flying east should "phase advance" their cycle by going to bed and waking up an hour or so earlier than usual; those flying west should attempt a "phase delay" by staying up an hour later and then sleeping a bit longer in the morning.

Disrupting one's biological rhythms can often cause more than the discomfort of fatigue; it can be hazardous to one's health. Studies have shown that people who frequently cross time zones, such as airline pilots and flight attendants, complain of a variety of physical ailments, including nausea, headaches, menstrual irregularities, and chronic sleep problems. They also report more marital and emotional troubles than people who follow regular schedules. Individuals whose jobs require working rotating shifts experience these problems and more: If they deal with hazardous machinery or materials, their very lives may be at risk.

Statistics show that most accidents caused by human error occur in the early morning and evening hours—between two a.m. and seven a.m., and two p.m. and five p.m. —when the body's inner rhythms are indicating a need for sleep. Nodding off at the job has been cited as a possible cause for the nuclear reactor accidents at Pennsylvania's Three Mile Island plant and at Chernobyl in the Soviet Union, and as a contributing factor in the space shuttle *Challenger* accident in 1986. Dr. Charles A. Czeisler, associate professor of medicine at Harvard University, has been involved in a series of studies of shift workers at atomic and automobile plants, in NASA's space shuttle program, and in

city police forces. According to Czeisler, 55 percent of the shift workers surveyed reported falling asleep at work. ''The two factors which cause the trouble,'' he explains, ''are disruption of the circadian cycle and sleep deprivation. Shiftworkers classically have to perform at a time when their brains are trying to put them to sleep.'' Typically, the workers move from shift to shift every week or two; thus, says Czeisler, their internal clocks never adjust to one shift before they must change again. In a pilot study of twenty-eight medical interns, Czeisler was surprised to find that during the course of one year more than 25 percent of the interns reported having fallen asleep while talking on the telephone, and 34 percent reported at least one actual or near-miss automobile accident because of sleepiness.

Some states have taken the initiative in limiting the working hours of their hospitals' interns and residents, and many businesses and municipalities are looking into the problems associated with shift work. Some are revamping schedules, rotating shifts no sooner than once every three

weeks, and giving workers several days off between shift changes. Since such a plan was implemented at a Utah chemical plant in 1980, managers have noted a decrease in illness, absenteeism, and the number of accidents. They have also reported a 20 percent boost in productivity.

Biological clocks possess an innate steadiness, then, yet are flexible enough, given the proper circumstances, to adjust themselves to accommodate such potentially disrupting situations as shift work. Does this mean there is some mysterious master clock that sets the pace for the body's various rhythms? For more than three decades, scientists have been searching for clues to this puzzle, and only recently have they identified a likely candidate—the suprachiasmatic nucleus, a small knot of tissue at the center of the brain.

Robert Y. Moore, chairman of the neurology department at the State University of New York in Stony Brook, was the first to suggest that this area of the brain set the body's tempos. He showed in laboratory experiments with rats that when the suprachiasmatic nucleus is destroyed, the rats' circadian patterns vanish. Although some scientists have disputed Moore's findings over the years, biologists working at the University of Virginia in Charlottesville in 1988 conducted a compelling study using golden hamsters, which normally follow a twenty-four-hour circadian cycle. Two new strains of hamsters with different cycles were bred for the experiment; they had daily rhythms of twenty and twenty-two hours. The scientists then destroyed the suprachiasmatic nuclei in the specially bred hamsters and in a normal control group, replacing them in each case with implants of different suprachiasmatic tissue taken from fetal hamsters known to have inherited either twenty, twenty-two, or twenty-four-hour circadian rhythms. After recovering from the transplant, hamsters from the three test groups exhibited entirely new rhythmic cycles, lengthening or shortening their days in step with the donor hamsters' circadian clocks.

While scientists are impressed by these results, they

Because of normal human sleep patterns, nighttime means danger in industrial settings such as the Three Mile Island nuclear power plant, shown here operating in 1986 even as cleanup work continues at Unit 2 (far left), site of a nuclear accident in 1979. The near disaster began in the early morning hours of March 28, when plant technicians, their natural sleep cycles out of kilter because of shift work, failed to recognize cooling-water loss caused by a stuck valve. They may have experienced "automatic behavior," a condition sometimes observed among overnight train drivers, who easily perform practiced tasks but may miss new information, even flashing danger signals.

caution that confirming the role the suprachiasmatic nucleus plays in humans is more difficult. Any such studies would have to take into account the other factors that affect human biological rhythms—external cues, or zeitgebers, and social cues, such as that provided by an alarm clock.

The interaction of these internal rhythms with external and social time prompts also plays a large part in an individual's psychological perception of time, as demonstrated by the case of Stefania Follini, who lost track of time during her isolation from time cues in the New Mexico cave. More typically, time, as perceived by the mind, may appear to drag or speed by, depending on such factors as where we are, what we are doing, and who we are with. Focusing intently on a task, for example—effectively shutting out all external and social time prompts—generally makes people unaware of the passage of time. One common example is that of rescue workers, for whom time is almost obliterated as they attempt to extricate or resuscitate accident victims. And surgeons in the operating room may lose all sense of time. When asked how he had physically endured a marathon twenty-four-hour operation to reattach the four severed fingers of an eighteen-year-old patient, a surgeon remarked, "I wasn't conscious of time." When a person is under such consuming circumstances, time appears to stand still.

Another possible explanation for how the mind senses time's passage is physiological. Some experts believe that an increase or decrease in an individual's sensory alertness mobilizes the brain to prepare a response and to screen out all distractions, causing a corresponding change in temporal perception. Feelings of intense fear, for example, can expand time, making a split second seem like an eternity. Survivors of automobile and airplane crashes often relate how they felt frozen in time during the brief seconds of terror that preceded their accidents. One who used those interminable moments to save himself from great injury was Albert von St. Gallen Heim, a University of Zurich geologist who in 1881 narrowly escaped death when he fell between the front and rear wheels of a wagon being pulled by galloping horses.

"For a fleeting moment, I was still able to hold on to the edge of the wagon," he later wrote. During that briefest of moments, Heim analyzed his predicament. He determined that his grip would not hold until the horses stopped, so he would have to let himself fall. He then decided that the best way to fall was to twist his body in such a way that he would land on his stomach. Finally, he told himself to tense his thigh muscles so his bones would be somewhat cushioned when the wagon wheel ran over his legs.

"I know quite clearly that I let myself fall only after these lightning-fast, wholly precise reflections," wrote Heim. Time had stood still, permitting him to think through a way to save himself. Only when Heim released his grip did time seem to move again. He fell, and the wagon rolled over him; but because he had "time" to prepare, his only injury was a bruise.

As Heim's story illustrates, the mind's ability to expand time during a life-threatening emergency may be a natural survival mechanism, a way to preserve the human species. It certainly helped save the life of Major Russ Stromberg, a Navy test pilot whose aircraft started losing power immediately after takeoff from a carrier in September 1980. In the space of just eight seconds, from the time of takeoff until the plane plunged into the sea, Stromberg had to decide how to save himself. "Everything went into slow

Machines That Alter Our Sense of Time

In technologically advanced societies, computers are increasingly common classroom equipment. Presumably, children will benefit from making an early acquaintance with machines that are likely to figure in the workplaces of the future. Some educators, however, charge that overexposure to computers, including the ubiquitous video games, can distort a child's perceptions of time.

Children come to their understanding of time in stages, initially learning to distinguish what comes first from what comes later. The next step is to notice duration. But computer use can disrupt the sense of duration; many children, caught up in the pace of work or play on a computer, lose track of time.

Speedy computers are also formidable competition for slower-moving humans. Once they become accustomed to rapid interaction between screen and fingertips, children may demonstrate impatience with people and noncomputer activities. Time spent with parents, siblings, or friends may seem to drag by; reading can become a frustratingly slow and seemingly inefficient way to gather information or pleasure.

Educators also worry that computer use may edge out natural experience of time, depriving children of the environmental cues and social activities that help develop their temporal sense. Nature's rhythms—the rising and setting of the sun, the passage of the seasons—may lose meaning to children attuned to the response time of computers; the organic tempo of human relationships might make scant impression on those whose primary interaction is with a machine.

Part of this concern arises from the speed with which computers have taken hold. The changes have happened so fast that most adults are ill equipped to teach children the necessary skills. Not burdened with old patterns of thought, children have an advantage in a computer age. Only time will tell how they use the new technology—and how they will be affected by it.

motion," the pilot recalled later. First, he tried to repower the engine. No luck. Then, with only precious moments remaining before the inevitable crash, he made plans to eject. "I couldn't tell by feel alone that I got a good grip on the handle of the ejection seat. I had to take time—even if there wasn't a lot of it—to look." Stromberg pulled the handle and ejected some thirty feet above the water, landing a safe distance from the plane's sinking debris.

Later it would take Stromberg forty-five minutes to describe what had passed through his mind during his plane's eight-second fall from the sky. While habit and experience obviously play a large part in making such split-second decisions as Stromberg's, some time researchers believe the pilot owes his life to his mind's ability to expand time—in this case to about 300 percent of normal time.

Scientists have just begun to unlock the mystery of how the mind stretches time. The key seems to lie in two sets of cells found in the frontal lobes of the brain. One set is clocklike, monitoring time moment by moment and enabling the brain to keep track of the present. The other set is involved in planning, judgment, and coordinating behavior. It helps the mind develop an awareness of time beyond the present: To make plans, the brain must be able to draw experience from the past and project actions into the future.

Scientists believe the close proximity of the planning and timing cells may be the cause of time distortion. When fear strikes, the brain becomes superalert, firing its cells at a terrific rate. The first to speed into action are the planning cells; all circuits light up as those cells search for a way to meet the perceived threat. The hyperactivity of the planning cells' circuits triggers similar activity in the timing cells, and they, too, go into overdrive. Instead of firing at the usual, slow ten impulses a second, the timing cells may double or even triple to a rate of twenty or thirty impulses a second. The brain is therefore racing along at two or three times its normal speed while clock time stays the same. As a result, clock time appears to drag.

Anticipation, as well as fear, can seem to slow the clock's hands. When you are waiting for something, notes

Robert Hicks, a psychiatry professor at the University of North Carolina Medical School, "you are in a state somewhat like that of an animal watching for a predator. You are totally focused on one event in the near future to the exclusion of everything else. You're not watching television or having a conversation. You're just focusing on time. The animal is completely focused on one future event, a specific threat. It's not going to look around for berries or wonder what's over the next hill. Both the person and the animal are in a state of hypervigilance, which is one where subjective time is accelerated." In other words, you believe that time is passing much faster than it actually is.

Although we often experience this type of time dilation, the most extreme example may be the seemingly endless days of a prisoner's life. As inmates adjust to jail life, they learn to fill their time with activities; counting the days until they are released only stretches time further. Sociologist Thomas Meisenhelder notes that "even the language used by prisoners reflects the feeling of effort associated with the passage of prison time." Serving a sentence is described in such phrases as "doing time," "marking time," "putting in" or "pulling" time. Writing about the same subject in a book entitled *Psychological Survival,* authors Stanley Cohen and Laurie Taylor observed that time for prisoners "is no longer a resource to be used, but rather an object to be contemplated." Prisoners are given extra time not as a reward but as punishment.

Perception of time's passage is also influenced by age. For young children, time goes by very slowly. The twenty-four hours before a much-loved event, such as Christmas, can seem an interminably long wait to a child. For an adult, however, those same twenty-four hours—indeed, the entire Christmas season—may appear to fly by all too quickly. Scientists offer a couple of explanations for why time creeps like a turtle for children and races like a hare for adults. One explanation is purely logarithmic: A year represents 20 percent of a five-year-old's life but only 2 percent of a fifty-year-old's life; thus, their perspectives on the experience of time's passage would naturally be different. Another, more

An inmate in the Arkansas state penitentiary fills empty hours with a daytime nap. Facing a future of monotonous sameness for the duration of the sentence, a prisoner typically attempts to speed time up by escaping into sleep and daydreams or by breaking up dull routines with spontaneous activities ranging from trips to the canteen to fights and even riots.

The slanting beams of the rising sun light the ghats of Varanasi, shrouded by mist from the sacred Ganges.

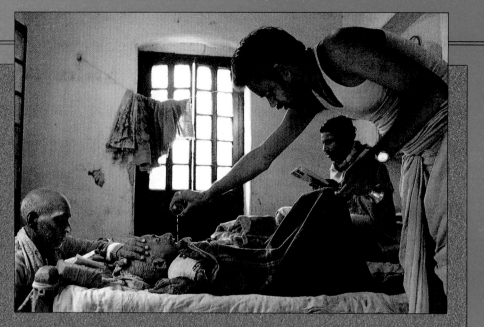

An attendant at the Mukti Bhavan (House of Salvation) trickles Ganges water into the mouth of a dying pilgrim. Any Hindu who expects to die soon may stay free at the hospice for fifteen days. Medicines are not permitted, and no food is provided; the rooms are barren cells. Comfort at the Mukti Bhavan comes from the atmosphere of peace and devotion, and the assurance that death will bring salvation.

Where Hindus Seek Escape from Time

To devout Hindus, earthly existence is an imprisoning chain of many lives, a journey of the soul through sequential incarnations that may take either animal or human form. The cycle of life, death, and rebirth is escaped only when an individual gains enlightenment and achieves reunion with the eternal, universal soul. A focal point in the quest for enlightenment is the holy city of Varanasi in northern India.

Believed to be the revered god Shiva's domicile, Varanasi is also the spot where a pillar of light, thought to reveal the secrets of transcendence, is said to have broken through the earth's crust. The city's religious importance is heightened by its location on the sacred Ganges River. Bathing in the Ganges at Varanasi, or drinking the river's water, is believed to wash away the sins of this and former lives.

Several miles of the riverbank are crowded with temples and ghats, steps or mud banks leading to the water. These religious sites attract hundreds of thousands of visitors annually. Those who can afford it arrange to die and be cremated in Varanasi, believing it will bring them deliverance; more than 30,000 bodies each year are consigned to flames on the ghats.

But assurance of mukti, or freedom from rebirth, is vouchsafed only to those who die within the city. At the moment of death, it is said, Shiva whispers to the dying person the taraka mantra, a verse that holds the key to enlightenment. Some people come to Varanasi in the hope of hearing this ultimate secret. In special ashrams and hospices, they await the end, confident that they will soon escape the confinement of earthly time and space.

A fortuneteller watches for customers in Macao, a tiny Portuguese enclave on the coast of southern China. His sidewalk stand is festooned with diagrams of palms and faces, marking the points he examines to interpret what the future might hold.

Searching for Signs of Things to Come

In modern Western cultures, the future is regarded as largely undefined. Growing from the present, its shape is considered discernible only to the extent that logic can be used to extrapolate trends. But a belief common in certain cultures is that the future has a predetermined shape that can be discovered, at least in its vague outlines, and used for guidance in daily living.

Foretelling takes many forms. In Tibet, fortunetellers may meditate, throw dice, or observe the behavior of birds to advise on important decisions. The diviners, typically Buddhist monks called lamas, may predict auspicious days for villagers to marry, to plant crops, or to plan military strategy; they might also choose the site for a new house or give insight into an individual's character and probable destiny.

Another popular fortunetelling method is palm reading, a technique developed some 3,000 years ago in China and India. Palmists scrutinize the lines of a hand to uncover such things as disposition, health tendencies, or future changes in one's chosen path.

The reading of sand pictures, another ancient means of tapping into the future, is practiced by several Native American tribes. Among the Navajo, tribal wise men trickle colored sand through their fingers, then interpret the resulting patterns. So magical are the pictures, believers say, that they must be erased, lest evildoers study them and learn the secrets they hold.

In parts of West Africa, villagers eager for advice rely upon a particular kind of spider believed to possess supernatural wisdom. Special fortunetelling cards, cut from flat, durable leaves, are left before a burrow inhabited by one of the large, hairy spiders. When the creature emerges, its roaming rearranges the cards. The diviner then interprets the new pattern and the way of the future is revealed.

intriguing explanation has to do with the body's biological rhythms. As people age and their metabolic rate slows down, some of their internal rhythms take on new beats. The sleep-wake cycle, for example, changes significantly; people past fifty tend to sleep less at night and nap more during the day than they did when they were younger. It may be that the biochemical processes in the brain that affect a person's sense of time also change with age, making time seem to speed by at an ever-increasing clip.

Another biological influence on an individual's time sense is body temperature. Neurophysiologist Hudson Hoagland, cofounder of the Worcester Foundation for Experimental Biology in Shrewsbury, Massachusetts, was the first to discover the connection between body temperature and time perception. He chanced upon it one day in 1933 while caring for his sick wife, who had a temperature of 104 degrees Fahrenheit. His wife asked him to drive to the drugstore for some medication. The trip took Hoagland only twenty minutes, but when he returned, his wife insisted he had been away for hours. Wondering if there was a correlation between his wife's temperature and her skewed perception of time, Hoagland asked her to count to sixty at a rate she thought approximated one number per second. She complied, reaching sixty well before a minute had passed. Hoagland had his wife repeat the experiment several times during her convalescence. He discovered that the higher his wife's temperature, the faster she counted. Studies by other scientists have confirmed his observations.

Rising and falling body temperature may also help explain why an individual's perception of time changes during the day. Studies have shown that late at night and early in the morning, when body temperature is low, time seems to speed up. Conversely, during the late afternoon and early evening hours, when body temperature is high, time appears to drag. Perhaps this is why commuters frequently report that rush-hour traffic in the afternoon is much more difficult to endure than the same congestion in the morning.

Psychologist William James observed in his 1922 book, *Principles of Psychology*, that there is a difference between

A Yemenite muezzin (above) calls Muslims to prayer, an activity that helps mark time in the Islamic world. Five times a day—sunrise, noon, afternoon, sunset, and evening—devout Muslims, like the Kuwaiti at right, stop wherever they are and kneel in prayer. The ritual also has a spatial obligation, requiring the faithful to turn toward the holy city of Mecca. In mosques, which are favored prayer sites, one wall is oriented so that anyone facing it is also facing Mecca.

البنك الاهلي الكويتي

ALAHLI BANK OF KUWAITKSC

الخزينة الليلية

NIGHT SAFE

the experience of time as it is going by—what he called the specious present—and one's reflection on that time after it has passed. "In general," James pointed out, "a time filled with varied and interesting experiences seems short in passing, but long as we look back. On the other hand, a tract of time empty of experiences seems long in passing, but in retrospect short." Although anyone incarcerated for a period of time would likely disagree, James believed this strange anomaly was due to the way time is conceptualized. People use events to define a particular span of time that occurred in the past. The more interesting the events, the more likely those events will be remembered and the "fuller" or "longer" that time will seem in retrospect.

How people perceive either past or present time is also a result of their culture. In modern Western societies, time stretches out along a horizontal path, one event following another, marching on into the future. From such a time line springs the Western notion of progress, that today is better than yesterday and that tomorrow will surpass today. Conversely, many non-Western cultures have a vertical, essentially static vision of time; for them, the events of the present are intertwined with those of the past and future. Tradition is valued as well as progress.

The Iroquois Indians of the American Northeast are an exceptional example of a culture with a vertical time perspective. When members of the Iroquois nation gather together to make a decision, they consider the wisdom of their ancestors and the needs of their future descendants as well as their current desires. Past, present, and future become part of a single time frame. As a result, the Iroquois are remarkably forward-looking, much more so than Westerners, who tend to be interested only in the immediate future. "Every decision we make relates to the welfare and well-being of the seventh generation to come," explains an Iroquois chief, "and that is the basis by which we make decisions in council. We consider: Will this be to the benefit of the seventh generation? This is a guideline."

Whether a culture has a vertical or horizontal perspective on time tends to determine how it organizes time. Cul-tures that take a horizontal view, most notably those of northern Europe, Japan, and the United States, prefer to schedule events one at a time. Because these cultures, which anthropologist Edward T. Hall calls monochronic, see time as running on a continuum, they find it easy to divide and compartmentalize time into separate units. Conversely, cultures with a vertical view of time, such as those of the Middle East, the Mediterranean, and South America, tend to have a more flexible approach to scheduling events. People from these cultures, which Hall calls polychronic, prefer to keep several activities going at once. Thus, the Navajo trader runs his business like a salon, with many people wandering in and out, discussing news, gossip, and personal affairs as well as conducting business transactions.

In monochronic cultures, the schedule becomes sacred. Trains and planes are expected to run on time. Appointments are made to the quarter hour. In Japan, where the business world is highly monochronic, working environments are tightly scheduled, from the workers' morning exercise program to their afternoon tea break. Even the traffic patterns are tightly controlled for maximum speed and efficiency. According to the Osaka Prefectural Police, a traffic-control system installed in that city in 1975 has reduced travel time by 17 percent and has saved more than 200 billion yen (about $850 million) in valuable time.

olychronic time, by contrast, is slower, more people-oriented. The process of completing a transaction, particularly the interaction of the people involved, is valued more than preset schedules. As a result, getting to an appointment on time is not taken as seriously. In Ecuador, for example, a woman on her way to an appointment may stop to talk to someone on the street and discover that a mutual friend is in the hospital. She immediately changes direction and heads for the hospital to visit her friend. Her first appointment must wait.

American psychology professor Robert Levine witnessed the difference between monochronic and polychronic cultures firsthand in the early 1980s, when he accepted a teaching position at a university in Niterói, Brazil, a city located across the bay from Rio de Janeiro. Levine's first class was scheduled to run from ten in the morning until noon. He started the class promptly at ten, but it was not until after eleven that all his students arrived. To Levine's surprise, although a few apologized for their tardiness, none seemed overly concerned.

Levine was equally bewildered when it came time for the class to end and only a few students left. The others stayed in their seats, asking questions or listening to the lingering discussion. Levine had never seen such behavior in an American classroom, where, he observed, students indicated class time was up by shuffling their books and wearing "strained expressions." Finally, sensing that the Brazilian students might continue the discussion for hours,

Levine excused himself from the room and the class ended.

Levine's Brazilian students came from a polychronic culture, where time is not perceived as having rigid boundaries. To them, the ten-to-twelve-o'clock time slot for Levine's class meant "late morning" rather than a precisely defined period of time. The experience so piqued the professor's interest in cultural perceptions of time that he initiated a study among Brazilian and American college students. When asked how they would define tardiness, the average Brazilian student answered that after thirty-three and a half minutes someone would be considered late; the American students felt that after nineteen minutes the person would be late. Brazilians were also more flexible concerning early arrivals—they allowed an average of fifty-four minutes before considering someone early, while the American stu-

The exterior scene blurs as it rushes toward the windows of a Japanese bullet train racing along at 120 miles per hour. Travel time on the Tokyo-Osaka run, formerly six and a half hours, has been halved by the high-speed trains. The concept of time as a scarce resource is not a modern fad in Japan; as early as the 1600s, rulers routinely exhorted their subjects not to waste it.

dents allowed only twenty-four minutes. Unforeseen circumstances was a common reason given by the Brazilians for arriving early or late to an appointment.

Levine and an associate then went on to study the pace of life in six countries—Japan, Taiwan, Indonesia, Italy, England, and the United States. They chose three indicators of time on which to base their comparison: the accuracy of each country's bank clocks, the speed at which pedestrians walked, and the average time it took a postal clerk to sell a single stamp. In each case they made observations in the nation's largest city and in a medium-size municipality.

Not surprisingly, Japan set the pace in every category. A survey of thirty bank clocks revealed that the Japanese timepieces were accurate to within thirty seconds of the time reported by the local telephone company. The United States placed second in this category, with Indonesian clocks being the least accurate—more than three minutes off the mark. The Japanese were also the fastest walkers, with the English placing second and the Indonesians finishing last. But the most telling difference between the cultures was witnessed in the selling of a postage stamp. Again, the Japanese completed the transaction in record time—twenty-five seconds—followed by the United States. The Italians came in last, taking an average of forty-seven seconds.

It was in Indonesia, however, that Levine had an interesting encounter with a polychronic culture. At the central post office in Jakarta, the professor was directed to a group of private vendors sitting outside who vied for his business by shouting "Hey, good stamps, mister!" At the main post office in the smaller city of Solo, he arrived on Friday after-

noon to find a volleyball game in progress outside; he was told business hours were over. When Levine next visited the office, he found the postal clerk was more interested in talking about America and his relatives in Cincinnati than in selling stamps. The five people lined up behind Levine waited patiently, he said. "Instead of complaining, they began paying attention to the conversation."

Unlike their polychronic counterparts, Westerners strongly believe in what psychologists call closure, the bringing of a task or activity to conclusion before beginning something else. Cultures with a vertical view of time, on the other hand, tend to be quite comfortable with putting aside an unfinished activity indefinitely. They have a very different notion about how events should be sequenced. In his 1983 book, *The Dance of Life,* Edward Hall recalls a confrontation between the Pueblo Indians and the government of the state of New Mexico during the late 1970s. Angry that the state had built a road on their land without compensating them for public access and right of way, Pueblo tribal leaders informed state officials that they would close down the road unless suitable restitution was made. Negotiations soon broke down, however, and the two sides remained silent on the issue for several years. Then suddenly, without warning, the Indians took action. They blockaded the road with a steel guardrail and attached a sign stating that they were exercising their right to close the road.

tate officials were bewildered. Why had the Indians acted now, after years of silence? To the Pueblo, who do not share the Western need for quick closure of events, it seemed quite natural to suspend action on a matter for several years before resolving it. Said one Pueblo leader, "I don't know why they were surprised. After all, those signs saying we were closing the road were stacked up against my house for a year and everybody saw them. What did they think those signs meant?"

Ideas about the order in which things should unfold is not the only dimension of time that differs among societies. Cultures also have their own unique ways of describing duration, the time during which something exists or lasts. Much of the world has now adopted the Western method of measuring duration, with the abstract and arbitrary units of seconds, minutes, and hours *(page 110).* In many traditional cultures, however, time is measured not by the ticking of the clock and abstract numbers but by concrete and often colorful descriptions of specific tasks or activities. As one scholar observed, the people of traditional cultures "don't tell you what time it is; they tell you what kind of time it is."

For example, a Oaxacan Indian of Mexico will refer to the distance between two villages as "a hat and a half" or "two hats," according to the time it takes to plait straw into a hat while walking. Or a Cross River native of West Africa, when asked how long it took a man to die, will reply, "The man died in less time than the time in which maize is not yet completely roasted," or less than fifteen minutes.

The concept of the week arose out of the almost universal need of various societies to schedule market days at regular intervals. Different cultures have given the week different durations, ranging from five to ten days. Even today, the Tiv people of central Nigeria recognize a five-day week, with each day named after the market that is held on that day. Instead of referring to a specific day as "Tuesday," the Tiv will call it "the day of the Sharwan market" or "Sharwan day." They also use the five-day cycle to describe time that has passed. For example, to indicate that he has been in a particular village for fifteen days, a Tiv man will say "I have been here three markets," meaning that the local market has been held three times since his arrival.

The human experience of time is thus a complex mix of the innate and the learned. To operate in modern society, individuals must coordinate three time worlds—the biological one they inherit; the psychological one they experience; and the cultural or social one they live in. Of these, the first remains sovereign—even if humans do not always perceive it as so. As the nineteenth-century French scientist Claude Bernard observed, the body will always have a relationship with the natural world that is "close and wise."

Taking the Measure of Time

Death is certain, only the hour is uncertain" reads the Latin inscription on the sixteenth-century sundial above. Although human beings have always been conscious of their mortality, this morbid message and the skull beneath it were likely inspired by the heightened awareness of time's passing that such timekeeping devices imparted.

For countless millennia, people had regarded time as a calm, ceaseless flow, and its passage was observed through the cycles of nature. People arose and retired with the sun and attuned their year to the changing seasons. But humans sought to understand time more fully and to control it, developing first the calendar to track the seasons and lunar phases, then such devices as the sundial and water clock to break time up into smaller increments.

As time-measurement implements became increasingly sophisticated, precise, and attainable, they gradually but markedly altered the way individuals regarded time. The advent of the mechanical clock, with its relentless march of measured moments, reinforced the notion of time as valuable and manageable—no longer a tide with which to swim effortlessly but a flow of limited quantity to be used to the fullest. Having declared their independence from the sun, people found themselves enslaved by a new master. The clock began to regulate most aspects of life, prodding and hurrying mortals with its chimes and ceaseless ticking. The following pages trace this evolution of time-measurement devices and how they have defined humankind's relationship with time.

These pages from a twelfth-century Mayan almanac chart the cycles of time associated with Venus (left) and Mercury (right). Mayan priests deemed it vital to know the planets' positions relative to each other and to other heavenly bodies plotted on Mayan calendars in order to determine the auspiciousness of any one day. Venus, depicted as the god of the evening star in the center drawing, was considered especially potent, since the planet was thought to influence the sun, rain, and war gods.

Calendars Based on Heavenly Movements

Early peoples took the first step toward measuring time when they moved beyond merely observing the rising and setting of the sun and began noting the length of the sun's shadows and its position in the sky during the changing seasons. By counting the number of sunsets, or days, that fell between successive solar events, such as from one summer or winter solstice to the next, early peoples were able to determine the approximate length of the year.

This knowledge helped farmers predict when seasons would begin and end and thus enabled them to plan the best times for planting or harvesting crops. These predictions achieved new heights of accuracy in Egypt about

6,000 years ago, when astronomer-priests noticed that once a year the bright star Sirius rose at dawn in direct line with the rising sun. Moreover, this so-called heliacal rising coincided with the yearly flooding of the Nile River, the event that deposited fertile soil along the riverbank and thus signaled the start of a new planting season. The priests used this knowledge to measure more precisely the length of the year, which they converted into a calendar consisting of twelve 30-day months followed by 5 feast days, for a total of 365 days.

Ordinary Egyptians—who viewed the sun, moon, and stars as gods to be feared, honored, and propitiated—had access to the measurement of the year

only through the skilled stargazers of the priesthood. Shrouding the calendar in mystery, the priests led others to believe that their intervention with the celestial powers ushered in and expelled the seasons and ensured the daily course of the sun. Indeed, the ability to predict lunar and solar eclipses and the annual flooding of the Nile must have seemed magical to those ignorant of astronomy. And this priestly monopoly was not unique to Egypt: There were several other ancient civilizations, including the early peoples of Europe and the Mayan Indians of Central America *(pages 34-36),* whose priests alone controlled time measurement, which was in turn ruled by the movements of the heavens.

Megalithic circles such as England's Stonehenge (below) are thought to have been employed by ancient priests as celestial observatories. Completed about 1100 BC, Stonehenge may have been used to predict the occurrence of solstices and lunar phases and eclipses—events that served as guideposts for marking off a year's time.

Seasonal scenes and zodiac signs adorn the calendar pages of (from left) March, June, September, and February in a 1409 book of hours.

Keeping Close Track of Dates

In Europe as elsewhere, the tracking of time gradually slipped from the exclusive grasp of the priestly caste into the public domain as civilization advanced. The calendar, which organized accumulations of days into weeks, months, and years, allowed an increasingly urban society to document events in the past, set plans for the future, and fix dates for rents and loan payments, as well as keep track of religious festivals.

By the medieval era, owning a calendar even became something of a status symbol: Aristocrats commissioned lavishly illustrated tomes called books of hours. These devotional works, which contained prayers to be said at appointed times, always began with

calendar pages for the twelve months. In addition to illustrations of some seasonal activity, the calendars frequently featured the sign of the zodiac appropriate to each month, since determining the astrologically correct occasion for various undertakings was an important aspect of time reckoning in medieval Europe.

The calendar followed by most Westerners in the Middle Ages was the one instituted by Julius Caesar in 46 BC. Because the Julian year did not correspond exactly with the length of the solar year, however, the calendar had long since fallen out of step with the seasons. In 1582, to correct the timing of movable feasts such as Easter, whose date is tied to the cycle of the

moon, Pope Gregory XIII imposed a new calendar that slashed ten days out of that year and readjusted the leap years. People all over Europe protested what they considered to be the loss of ten days from their lives, but the new calendar prevailed, at least in most Catholic countries. England and the American colonies refused to adopt the Gregorian calendar until 1752; Russia held out until 1918. In the meantime, calendars remained a political tool of sorts. In 1792, for instance, the French revolutionary government imposed a new calendar that excluded all Christian elements; thirteen years later, however, under the rule of Napoleon Bonaparte, it was abandoned in favor of the Gregorian model.

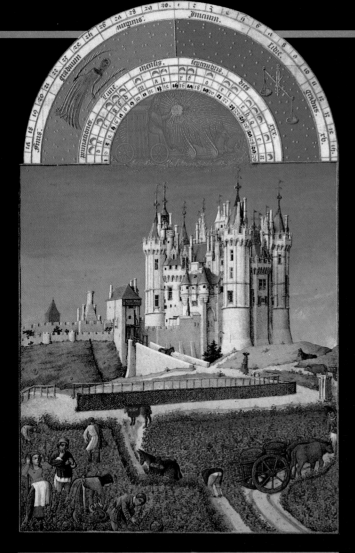

Cycus & pact. An̄ Cor 1582	Lfæ Uni- cales		Dies mon- sis,	OCTOBER Cui defunt decē dies d correctiōe Anni Solatis
xxij	A	kal	1	Remigij Epi & Côfeſſ.
xxi	b	vi	2	
xc	c	v	3	
xix	d	iiii. No.	4	Franciſci Confeſſ. dupl.
viij	a	Idibus.	15	Dionyſij, Ruſtici, & Eleu therij mart. ſemid. cū cōm. S. Marci papæ, & conf. & SS Sergij, Bac chi, Marcelli, & Apu leij martyrum.
vij	b	xvij	16	Caliſti papæ & m̄. ſemi
vj	c	xvi	17	
v	d	xv	18	Lucæ Euãgeliſtæ. dup.
iiij	e	xiiij	19	
iij	f	xiij	20	
ij	g	xij	21	Hilarionis Abbatis . & cōm SS. Vrſulæ & ſo. Virginum, & mar.
j	A	xj	22	
✠	b	x	23	
xxix	c	ix	24	
xxviij	d	viij	25	Chryſanthi & Dariæ m̄.
xxvij	e	vij	26	Euariſti Papæ & mar. Vigilia.
xxvj	f	vj	27	
25	xxv	g	v	28	Simon's & Iudæ Apoſt. lorum. duplex.
xxiiij	A	iiij	29	
xxiij	b	iij	30	
xxij	c	Prid.	31	Vigilia

In his reform of the Julian calendar in 1582, Gregory XIII cut ten days from one month, ordaining that October 4 be followed by the 15th, as shown on a page from the papal bull at left. Refined with amendments such as leap seconds added about every two years, the Gregorian calendar is still widely used.

A 1790s perpetual calendar (below) contrasts the Gregorian calendar with the French revolutionary version. The French divided months into three 10-day weeks, named the months after the seasons, and assigned numerals to the weekdays. The month names are compared at the lower corners, the day names in the list at center.

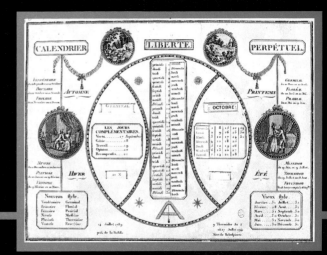

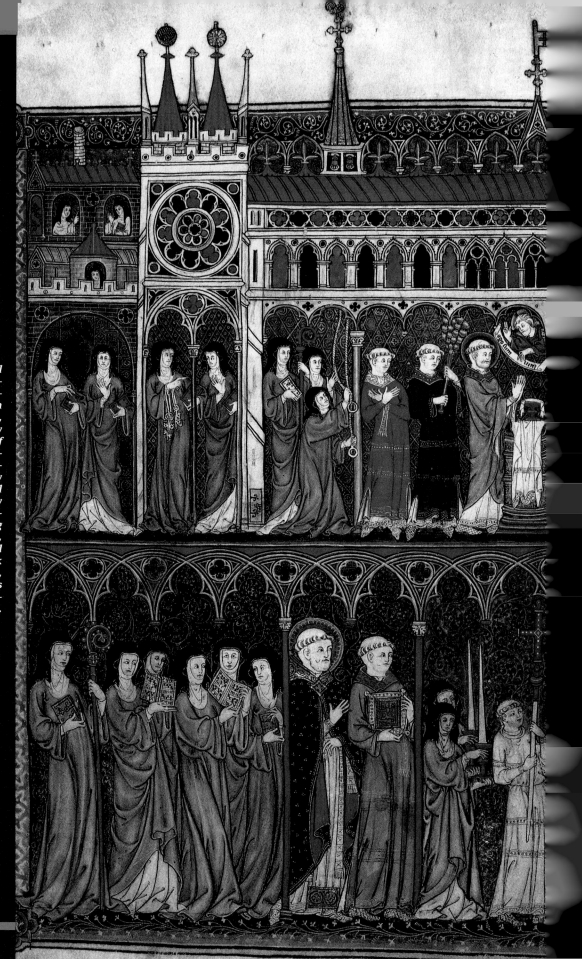

As the sacristan pulls the bell ropes to signal the time, cloistered monks and nuns celebrate Mass in this illustration from a French manuscript, dating to about 1300. Nearly every act in the daily life of medieval monasteries and nunneries—worship, meals, bedtime, rising—was regulated by bells sounding the canonical hours, the times established by the church for various devotions. Responsibility for ringing the bells lay with the sacristan, who, before the mechanical clock was invented, kept track of the hours with water clocks, sandglasses, or candle clocks —which melted away calibrated marks as they burned.

Marking the Hours

Just as the sun's seasonal travels across the heavens inspired the creation of the calendar, so its daily journey provided the first means of dividing the day into segments. By tracking the movement of the shadow cast by a stick that was set upright in the ground, people could determine what portion of the day had passed and, fairly exactly, how much work they could accomplish before darkness.

By about the fifteenth century BC, the Egyptians had refined this time-keeping concept into a sundial, which charted the progression of the sun's shadow over a circular base divided into twelve sections, or hours. Although the length of each hour varied, being dictated by the cycles of the sun and the seasons, the device nonetheless divided time into arbitrary increments—a step that led time measurement, however slightly, away from natural forces and into human hands.

In their sunny climate, the Egyptians were well served by shadow clocks. But they also wanted to measure time at night. For this, they invented the water clock. Particularly useful for marking small intervals of time, the water clock worked by measuring the flow of water from one vessel into another. While the Egyptian elite employed the clocks mainly for private use, the Greeks and Romans used them—along with sundials—to keep public track of time. Elaborate water clocks with floats and dials were erected in public squares so that all citizens might know the hour.

Along with this increased attention to the inexorable passing of time came an increasing awareness of mortality. One Roman patrician reportedly had not only a water clock in his dining room but "a uniformed trumpeter to keep telling him how much of his life is lost and gone." Later, the sandglass reinforced this image of impending death: In England, it was often placed in coffins to show that life's time had run out.

Shaped like a half-bowl, the hemicyclium at left, a refinement of early sundials, divided daylight into twelve equal hours. Its gnomon—the part of the sundial that casts a shadow—is missing but must have extended horizontally from the top center of the bowl. The shadow path within the bowl mirrored the sun's arc through the sky.

This reconstruction of a sinking-bowl water clock (below) consists of a bowl with a hole in its bottom placed inside a water-filled reservoir. As the bowl filled, it would sink. When the vessel was completely submerged, an attendant would sound a signal and then re-float the bowl.

This carved sandglass, crafted in France or Italy during the 1700s, measures an hour; its bulbs indicate the quarter-hours. Thought to have been invented by a monk sometime in the eighth century, sandglasses have been used to time sermons, lectures, work hours, the speed of a ship, and the cooking of eggs.

*Completed in 1518, the soaring clock tower of Antwerp Cathedral stands
as a monument to the evolving time-consciousness of the era. Although the earliest public clocks
merely tolled the hours, later models such as this also displayed the time on a dial.*

The Mechanical Revolution

Nowhere was timekeeping more important than in medieval monasteries, where strict adherence to the canonical hours of devotion defined the way of life. But the devices used to mark time's flow were neither dependable nor convenient: Water clocks froze in cold weather, and sandglasses and candle clocks had to be monitored diligently. In AD 996, spurred perhaps by the shortcomings of those ancient instruments, Pope Sylvester was credited with creating a wheel-and-weight-driven clock designed to tell the sacristan when to ring the bell. Although it was not commonly used until much later, the timepiece was the first of many mechanical clocks that would revolutionize how humans not only measured time but thought of it.

By the early 1300s, weight-driven clocks that sounded a bell were used in monasteries throughout Europe. Around the same time, larger turret clocks were placed in monastic towers, where they tolled the hours for the surrounding community and for the monks. Because the sun's path defined the beginning and end of the day, the earliest striking clocks were designed to remain silent at night. But by the mid-1300s, the mechanical logic of the machine, which could run without supervision throughout the night, led inevitably to the incorporation of the hours of sunlight and darkness into a single day composed of twenty-four equal hours.

Over the next century, nearly every European town of size installed a clock in the tower of its church or town hall, reflecting the growing demand for accurate timekeeping outside the Church. Influenced by the regular announcement of the hours and later the quarter-hours, people quickly ceased regarding time as an abundant, cyclic flow and instead saw it as an accumulation of discrete measured moments—a precious commodity that could be spent wisely or squandered. And as portable clocks and watches became readily available in the 1600s, life began to be scheduled by these mechanical usurpers of the sun.

An ingenious tabletop alarm clock (above), made in Germany around 1600, awoke its owner with an exploding gunpowder charge. A wheel-lock mechanism, automatically tripped at the set time, produced a spark that ignited the charge and lighted a candle attached to the clock.

Representative of the lavish timepieces owned by the status conscious for many generations, the clock and watch at right were made for the duke of Orléans in 1835. When placed in its holder atop the clock, the pocket watch was wound and set by the larger timepiece to the matching time.

Life by the Clock

Although by the mid-seventeenth century portable clocks were used for such purposes as measuring the craftsman's hours or fixing the times for appointments, the devices were still relatively inaccurate. It was rare indeed to find any two clocks that kept the same time. But toward the end of the century, the incorporation of the pendulum and balance spring into mechanical clocks reduced their variations from about fifteen minutes per day to less than one minute. And by the early 1700s, accurate, portable timepieces were inexpensive enough for almost everyone to own.

But while clocks and watches now kept reliable hours by which all classes of men and women could coordinate their activities, the time kept in one town often varied from that observed in a neighboring town. This minor annoyance became a question of public safety in the 1800s, with the introduction of the railroads. To ensure that schedules were met and collisions avoided, standard time zones were imposed to coordinate time among different regions.

Another product of the Industrial Revolution, mass-production factories, also operated strictly by the clock. The design of the machines and the allocation and nature of the jobs—all mined by time charts and stopw es—meant that workers had to at the factory at regular fixed tir Workdays were no longer define tasks completed but by the num hours on the job. Human time v controlled by the clock.

Indeed, in the modern world, dependence on calendars, preci clocks, and human-contrived tir zones, most men and women h lost touch with the natural pher that once defined the flow of tir the universe. Instead, few c make any decisions—weigh trivial, business or persona without consulting our ubic dictator, the clock.

To ensure that they were not cheated of a worker's time—by now a valuable commodity—nineteenth-century factory owners installed clocks that logged employees' arrival and departure times on cards the workers punched into the machine (right). Many such clocks worked to the employer's advantage: Clocking in a minute late forfeited an hour's pay.

In 1884, twenty-seven nation agreed upon an internationa time-zone system (map, belo that established the meridia passing through Greenwich, England, as the zero point fo time measuring. Now the un versal standard for timekeep ing, the system ensures that minutes and seconds are the same on clocks the world ov only the hour varies.

*Brokers at the Japanese stock exchange anxiously ply their trade under
the prominently displayed time—a visual reminder that every minute counts in today's international
financial markets, where fortunes are made or lost in the blink of an eye.*

CHAPTER 4

Strange Times, Strange Places

hortly before 9:00 on the morning of October 3, 1963, Mrs. Coleen Buterbaugh, a college secretary in Lincoln, Nebraska, stepped through a familiar office door and into what seemed to her a frightening dislocation of time. As Buterbaugh later recounted the story, she had been carrying a message to a professor across the campus of Nebraska Wesleyan University when the strange event occurred. It was a walk that she had taken hundreds of times, leading her past the college's recently completed Willard House as well as several other campus landmarks on the way to her destination, the old C. C. White building.

On reaching the building, she entered from the front and passed down a long hall that led to a suite of offices at the rear. As she walked, Buterbaugh heard the familiar sounds of students on their way to class and snatches of music drifting from the rooms reserved for music practice. In particular, she heard the sound of a marimba.

All at once, upon entering the offices where she expected to find the professor she was seeking, Buterbaugh was nearly overwhelmed by a powerful, musty odor. "As I first walked into the room everything was quite normal," she later recalled. "About four steps into the room was when the strong odor hit me. When I say strong odor, I mean the kind that simply stops you in your tracks and almost chokes you. I was looking down at the floor, as one often does when walking, and as soon as that odor stopped me I felt that there was someone in the room with me."

Glancing up, Buterbaugh saw the figure of an unfamiliar woman straining to reach the upper shelves of an old music cabinet. Although the figure's back was turned, Buterbaugh could see that the woman was very tall and thin and that she wore her black hair in an old-fashioned bouffant style. Her clothes, including a long shirtwaisted dress, also appeared to be some thirty years out of date. At that moment, Buterbaugh became aware that she could no longer hear the familiar noises from the hallway. There was no sound of a marimba. Everything around her had grown deathly quiet, as though the rest of the world had simply faded away. In the stillness, she stood transfixed by the figure before her.

"She never moved," Buterbaugh reported. "She had her back to me, reaching up into one of the shelves of the cabinet with her right hand, and standing perfectly still. She was not transparent, and yet I knew she wasn't real. While I was looking at her she just faded away—not parts of her body one at a time, but her whole body all at once."

The strange encounter did not end with the figure's disappearance. "Up until the time she faded away I was not aware of anyone else being in the suite of rooms," Buterbaugh continued. "But just about the time of her fading out I felt as though I still was not alone. To my left was a desk and I had a feeling there was a man sitting at that desk. I turned around and saw no one, but I still felt his presence. When that feeling of his presence left I have no idea, because it was then, when I looked out the window behind that desk, that I got frightened and left the room. When I looked out that window there wasn't one modern thing out there. The street, which is less than a half block away from the building, was not even there and neither was the new Willard House. That was when I realized that these people were not in my time, but that I was back in their time."

Stunned and frightened, Buterbaugh fled the room. "It was not until I was back out in the hall that I again heard the familiar noises," she said. "This must have all taken place in a few seconds because the girls that were going into class as I entered the room were still going in, and someone was still playing the marimba."

Buterbaugh's case was investigated and reported by two respected psychical researchers, Gardner Murphy and Herbert L. Klemme. One of them, Murphy, was a distin-

guished psychologist with degrees from Yale, Harvard, and Columbia, who for many years was chairman of the Department of Psychology at the City College of New York and later director of research for the Menninger Foundation. He nurtured a lifelong interest in the paranormal and served as president of the American Society for Psychical Research before he died in 1979.

If it occurred as described, Buterbaugh's peculiar experience would be classified by students of alleged paranormal phenomena as a time slip—a supposed shift, usually sudden and short-lived, into a period of time other than the present. The most striking feature reported by individuals who claim to have undergone this experience is the apparent normality of the anachronistic world they enter. The people—or images of people—they have allegedly encountered there are not perceived as insubstantial wraiths but seem, at least momentarily, as solid as the observer. In fact, paranormal researchers say it is not uncommon for persons experiencing a time slip to assume they have blundered into a costume party or the filming of a movie.

In many ways, Buterbaugh's account resembles the more familiar reports of ghost sightings and phantom encounters. She did not begin to think she had slipped into another time, she said, until she happened to look out of the window and saw an unfamiliar scene that could have belonged only to the past. As soon as she had recovered her composure, Buterbaugh told her colleagues what had occurred. The figure she described sounded oddly familiar to some of the older staff members. Decades earlier, that office had been occupied by a woman named

Clarissa Mills, a lecturer in music theory and piano. Buterbaugh's description of the mysterious apparition, they thought, resembled Clarissa Mills, a woman she had never met. At once Buterbaugh hunted through some old college yearbooks. She said that when she located a picture of Miss Mills, she was struck by the similarity. Although the figure in the office had kept its back turned, the woman in the photograph wore her black hair in the same bouffant style. Miss Mills's clothing in the picture, too, matched the period and style of that worn by the mysterious visitor.

Further digging revealed an even stranger twist. Early one morning in 1936, Miss Mills had struggled through a bitter wind to reach the office building. Apparently the exertion had been too much for her. She died there shortly before 9:00 a.m.—the precise time of day that Buterbaugh believed she saw her twenty-seven years later.

Time slips like this one—if they actually occur—call into question our traditional understanding of time. They suggest, for instance, that the past is not over or finished but in some way still exists or at least has somehow left a record of itself that can be perceived with the senses. Indeed, if the past and the present are disrupted and become juxtaposed or intermingled in a case like Buterbaugh's, which time is "real"—that of the present-day observer or that of a person who exists in the past relative to the present-day observer? Is it possible that the apparition of Clarissa Mills, going about her business in the university office, could have turned and seen Buterbaugh? Many accounts of reported time slips involve conversations and other interactions between those who claim to find themselves in the past and the people they say they encounter there.

In fact, the simple, linear view of time embraced by most people in the West—that the present is here and now, the past is finished, and the future is yet to come—has never been accepted by many Eastern cultures. They see time as a forever-renewing cycle rather than as a straight progression. And now the popular, straightforward notion of time has also been challenged by science. Scientists have learned that there are physical phenomena at both the sub-

When Eileen Garrett allegedly looked into the past, it was often as a medium whose body appeared to be occupied by a spirit. At times during such trances, she would moan as if in pain and thrash violently, throwing herself to the ground. For all her theatricality, Garrett convinced many witnesses of her clairvoyance, and she devoted the greater part of her life to furthering the scientific understanding of such phenomena. In one test of her alleged abilities, she identified hidden objects, such as the Babylonian tablet at left. She described not only this artifact but also the secretary who had packaged it weeks before the test.

Reports of the Clairvoyant Process

During long careers scrutinized by skeptical scientists, reputed psychics Gerard Croiset and Peter Hurkos of Holland and Irishwoman Eileen Garrett, all of whom are now deceased, submitted to extensive laboratory testing. The results were far from definitive, but they seemed to display clairvoyance on a number of occasions—most perplexing, a seeming knack for perceiving the future and peering into the past.

Powers of precognition and retrocognition, as these clairvoyant sensitivities are called, have also been attributed to Alex Tanous, a current-day psychic of Lebanese-American extraction. Like Hurkos and Croiset, Tanous has at times worked with police in solving crimes and finding missing persons.

He claims that he mentally projects himself back in time to the point where a suspect or missing person was last seen and then witnesses the course of past events as they unfold.

Like his three predecessors, Tanous does not claim to understand how his powers work. But his experiences differ from theirs, as the descriptions here indicate. American parapsychologist Lawrence LeShan once noted that psychics are somewhat like religious mystics in that they enter an altered state of consciousness in which time and space seem to have little relevance. Eileen Garrett concurred with this assessment. "In the ultimate nature of the universe," she said, "there are no divisions in time and space."

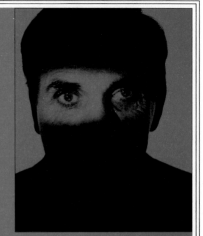

Alex Tanous (above) claims that his psychic messages arrive without fanfare—as just another series of words and pictures that drift into his mind. Tanous puts himself in a receptive state by staring at bright lights or walking in the sunshine. Peter Hurkos (below) likened his own extrasensory experiences to the flood of images on a television screen. He gained notoriety in a failed attempt to identify the Boston Strangler.

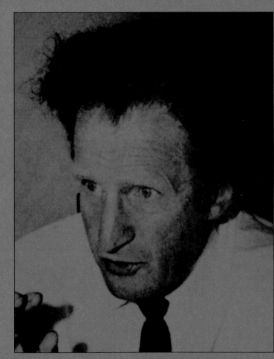

Like Garrett, Gerard Croiset (above) was intensely curious about the source of his peculiar insights. He described the experience of looking forward or backward in time as beginning with a vision of an amorphous field of powder that would arrange itself in dots and lines. As the vision progressed, the lines would form shapes and depict scenes in three dimensions. He had a bizarre aptitude for gazing across an empty meeting hall and telling of the people who would sit in each seat, days or even weeks hence.

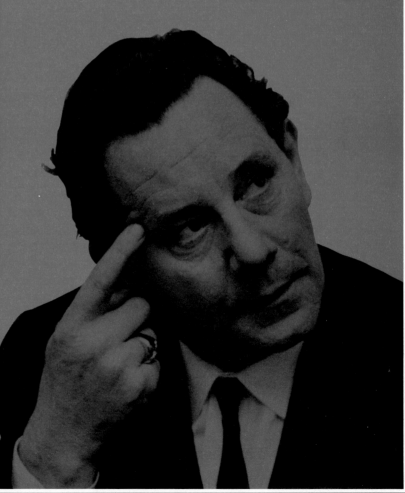

atomic and the cosmic levels that can be accounted for only if time is not absolute but relative, and not a separate dimension but part of a space-time continuum that can be bent, speeded up, slowed down, and perhaps, halted. Some ideas that were once dismissed as ridiculous—such as time travel—are now discussed as being theoretically possible *(pages 71-77)*. Physicists have even explored the hypothesis that time can actually be reversed. Nature, they have discovered, is indifferent to the direction of time. One can substitute a minus value for time in any basic law of physics and the law will still describe something that is a possible occurrence in nature.

To the dismay of many scientists—in whom respect for the scientific method and for high standards of proof tends to encourage an appropriately conservative attitude—these discoveries raise fascinating possibilities for people who like to speculate about so-called paranormal phenomena. Mysterious appearances and vanishings, uncanny coincidences, reported sightings of unidentified flying objects, even the age-old practice of dowsing to find water or mineral deposits—all these take on new significance for paranormal researchers when considered in the context of what scientists have learned about space and time. No less familiar a phenomenon than déjà vu—the strange feeling that you have experienced something before, when in fact you have just seen it, heard it, or done it for the first time—is now being reexamined in terms of the new concept of time.

The questions proliferate as rapidly as the possibilities. For instance, if the past and even the future are somehow accessible through time slips, retrocognition, or precognition, can the observer from the present actually change the past or future?

In one instance, a young woman who worked as a maid in a castle in Wales said that she caught an unexpected glimpse of a future accident but found herself unable to prevent it. According to the woman, she was standing in the castle's serving pantry one afternoon when she saw the senior maid drop a jug of melted chocolate while trying to hand it to a boy. The jug smashed on the floor and the choc-

olate spread into a brown, amoeba-shaped stain. The observer said that as she stared at the broken jug, the scene melted away and then started over again, like a loop of film run twice. This time, the woman tried to prevent the mishap. She shouted a warning to leave the jug alone, but the senior maid picked it up and broke it again in exactly the same way. "They said if I hadn't shouted," she recalled ruefully, "the whole thing wouldn't have happened."

Examples of reported precognition are sometimes more significant than kitchen spills. Danah Zohar, an American who earned her bachelor of science degree in physics and philosophy at the Massachusetts Institute of Technology, tells the story of a medical student in London, where Zohar settled to follow a writing career. Zohar claims that the student—to whom she gives the pseudonym of John Peters—was taking a written exam in 1956 when he came to a question about the biochemical steps by which the body makes fatty acids. In his answer, Peters described an important experiment, which he said had proved that the synthesis of a molecule called acetyl-coenzyme A, or acetyl-CoA, was the primary step in the creation of fatty acids.

ccording to the story, Peters's professor praised him for his imaginative answer but pointed out that no such experiment had ever been done and that acetyl-CoA was still just one of several possible molecules whose creation might be the starting point for the process. Peters was floored. "But you described this experiment in your lecture during term," he told the professor. "I've got it written here in my notes." The professor stood his ground: He had never described such an experiment, because the experiment had never been done.

Peters, who could not believe he had hallucinated the lecture, said that he eventually put the matter in the back of his mind. But eight years later, while reading a medical journal, he was sharply reminded of the experience. For there, in an article on the synthesis of fatty acids, was an account of an experiment conclusively proving that acetyl-CoA was the building block with which the process began.

The experiment, he said, was exactly as he had described it on the exam, except that the article stated it had only just been done—for the first time.

Apparently, however, supposed glimpses of the future occur in dreams more frequently than when the observer is awake. Often the subject matter is insignificant. One man, a British radio engineer, was said to be troubled by a dream that a sparrow hawk was perched on his right shoulder. The dream was remarkably vivid; the man claimed he could actually feel the bird's sharp claws digging through his jacket. The next morning, his landlord came by offering some bits of scrap that could be burned in the fireplace. One of the items he left was a stuffed sparrow hawk mounted on a wooden base. Later, one of the engineer's colleagues was looking through the box and chanced upon the bird. He removed it from its base, came quietly up behind the engineer, and playfully pressed the bird down on the right shoulder of his friend's jacket. The engineer later said that as he felt the claws dig into his shoulder, he realized with a shock that his dream had come strangely true.

But dreams, too, can be about serious subjects. Cambridge astrophysicist and science reporter John Gribbin cites the example of a man named Malcolm Bessent, who "in the course of one night" in late 1969 was said to have dreamed forecasts of a disaster befalling a tanker owned by Aristotle Onassis within six months, the death of Charles de Gaulle within a year, and a new British government coming to power in the summer of 1970. The dreams, Gribbin says, all came true on schedule.

British author J. B. Priestley, who was fascinated with the psychic aspects of time, told the following story of a dream that was very important indeed for the dreamer, a butler named John Lee. John Lee, said Priestley, was convicted of murdering his elderly female employer and was sentenced to death by the rope in 1885. In England's Exeter prison on the night before he was to be executed, Lee dreamed that he stood long minutes waiting on the gallows, the noose around his neck, because the device that was meant to open the trapdoor beneath his feet would not

function. He dreamed he was finally returned to his cell, by a route that took him through areas of the prison he had never seen in waking life.

wo prison officers stood watch over him all night, and in the morning he told them about his "very singular and strange dream." The officers in turn informed the prison governor about the dream; thus three officials could attest to his claims of precognition when the attempt to hang him that day proceeded just as he had dreamed. The trapdoor would not open, the execution was called off, and Lee was returned to his cell through rooms that he had accurately described based on the dream. His sentence was commuted to life imprisonment; after some fifteen years he was released and went to America, where he survived until about 1940.

John Gribbin sees a direct relationship between reported incidents of precognition, awake or dreaming, and science's new understanding of time and space. He notes, for instance, that time does not move forward at the steady pace indicated by our clocks and calendars but can be warped and distorted, with the end result being a relative value depending on the observer's point of reference. He points out that supercollapsed objects in space—such as blacks holes are thought to be—can negate time altogether, making it stand still in their vicinity. Furthermore, Gribbin says, some physicists claim to have detected particles that actually travel backward in time.

For Gribbin, these anomalies bear directly on the human mind. "The evidence that some form of ill-understood mental process is able to short-circuit the normal 'causal' flow of time, and that this process works most effectively in providing some of us with 'precognitive' dreams, is now compelling," he writes in his book *Timewarps*, published in 1978. In some cases, he maintains, these precognitive glimpses of the future may represent mysterious resonances between individual minds, separated in time, but still able to swap detailed information about specific events. These resonances may one day unlock tremendous mysteries. Gribbin speculates, for example, that humankind's obsession with extraterrestrial travel and contact with other beings in space may represent a precognitive glimpse of the day when we will actually make such contact.

Danah Zohar, in her 1983 book *Through the Time Barrier,* concedes that scientists have yet to gather much in the way of hard evidence concerning precognitive phenomena. Like Gribbin, however, she believes that the human mind may one day reveal the nature of time. "If it does exist," she notes, "precognition would not be the only one of our faculties which still eludes scientific rigor. For all of their conceptual advances, scientists still know very little about consciousness or the human brain. The full mechanics of ordinary perception, the functioning of long-term memory and the relation between 'mind' and 'body' are still beyond the pale of scientific explanation."

The difficulty in explaining precognition, Zohar believes, is that it forces us to make physical sense of a future already in existence. Zohar says there are two interpretations of precognitive events: the traditional view that the observer is foreseeing an actual event that has not happened yet and a more modern notion that the event being perceived is a future possibility that may or may not happen. Both approaches are consistent with modern physics, but the second interpretation has a certain appeal because it allows us to believe in free will.

In 1960, Dr. Ninian Marshall, a young psychiatrist who had earlier proposed a quantum-mechanical basis for telepathy, presented a theory that attempted to explain how precognition could work in physical terms. Marshall suggested that disturbed subatomic particles, such as electrons, throw out "feelers" toward their futures.

The Sound (and Sight) of Music

For those whose appreciation of music exists only on an auditory level, the work of Geoffrey Hodson may seem fantastic. In the early 1930s, Hodson, an alleged clairvoyant and lecturer on theosophy, began researching the "different effects produced by the performance of music upon the adjacent matter of the super-physical worlds." That is, he listened to music and used his supposed powers to describe what it "looked" like. He reasoned that variations in combinations of tones, lengths of notes, and rhythms would produce various mental shapes and patterns peculiar to each piece of music.

His verbal descriptions of a number of pieces were recorded while the music was played; artists then used them as the basis for paintings. Handel's "The Harmonious Blacksmith" inspired a lily-like painting *(far left),* whereas the majestic "Hallelujah Chorus" from the *Messiah* was depicted by a more complex composition *(center).* A solid, broad-based "tree" *(left)* represented the somber tones of Bach's Prelude in C-sharp Minor. Bizarre as Hodson's work may seem, some think it might provide a basis for a greater sensitivity to music. Others say that many people already respond to music with interior visions and that the love of music is a personal experience unlikely to be affected by his ideas.

These feelers, called virtual transitions, act out all the futures possible for the particle—that is, they simultaneously explore all the possible energy states that the particle might next assume and then "decide" which one to choose. Precognition, said Marshall, might be explained similarly if the brain could somehow tune in on these dips into the future. He theorized that quantum events might band together in a pattern, allowing the brain to magnify them into a recognizable human event. He called these patterning and magnifying processes "resonance phenomena" and compared them to tuning forks or windowpanes vibrating in harmony with a passing train. Marshall's theory is a simple quantum-mechanical explanation of how something can be "seen" that has not yet happened. His emphasis on precognition as a vision of possible future events also helps to explain the randomness and imprecision of precognitive dreams.

American physicist Gerald Feinberg expanded on these ideas in a paper entitled "The Remembrance of Things Future," which he presented in 1974 at the International Congress on Quantum Physics and Parapsychology in Geneva. By comparing the similarities between precognition and short-term memory, Feinberg theorized that the brain may receive information from the future as well as the past. Short-term memory, he explained, is a processing mechanism in which bits of information are retained for a few minutes before they are permanently recorded or lost forever. Precognition, he suggested, might be pictured as short-term memory in reverse. If it received stimuli from the future, the brain might, in a sense, "remember" events that had not yet occurred. Feinberg based his speculation on a symmetry in the equations of electromagnetism that suggests, in theory at least, that it should be possible to receive information from the future as well as the past.

Still another kind of physical explanation of precognition is the so-called observation theory, espoused by American physicist Evan Harris Walker in 1974. Walker's ideas, like those of Marshall and Feinberg, take quantum mechanics as a starting point to address the question of how possibility becomes actuality. His observation theory states that any sequence of events is possible until we examine a phenomenon with measuring instruments, at which point one possibility becomes actual. The conscious act of foreseeing a future event has the effect of creating—retroactively—the event foreseen. In recent years, highly technical variations on Walker's theory have dominated the discussion of how precognition might work.

The common-sense view of time is based on causality: An event must always happen after the thing that caused it—a glass breaks after it is dropped on the floor, a baseball is hit after the bat is swung. If causality is a law of nature, it would appear to make precognition impossible. But modern physics seems to break through this barrier, forcing even the greatest scientists to revise their beliefs. In the 1920s, the famed physicist Albert Einstein denied firmly that time travel might be theoretically possible. Two decades later, however, in reviewing a paper on relativity, he admitted that on a cosmic scale it is merely "a matter of convention to say that A precedes B rather than vice versa."

Scientists and nonscientists have also speculated that coincidences, like apparently precognitive events, may be related to the physical laws of time and space. They maintain that coincidences, at least some coincidences, are not purely random or accidental as traditionally thought.

Certainly the weird experience of a Frenchman named Deschamps strains the limits of ordinary coincidence. The story, related by the French astronomer and philosopher Camille Flammarion, begins when Deschamps, as a boy in Orléans, was given a piece of plum pudding by an old family friend named de Fortgibu. Ten years later, M. Deschamps saw a plum pudding in a Paris restaurant and ordered some for himself. His waiter informed him that all of the pudding had already been ordered—by M. de Fortgibu. Many years later, the story took an even stranger turn. While attending a private party, Deschamps was invited to share in a special plum pudding. As he was eating, he remarked that the only thing missing was his old friend de Fortgibu. At that moment, the door to the room flew open and a very elderly man stumbled in. It was de Fortgibu, who had gotten the wrong address and burst in on the party by mistake.

In 1919, an Austrian biologist named Paul Kammerer wrote that coincidences did not happen randomly in time and space but were grouped together in series. He called this effect the law of seriality. It was the famous Swiss psychoanalyst Carl Jung, however, who probably attracted the most-lasting attention for work on the subject of coincidences, particularly for his efforts to explain them in terms of what he called synchronicity.

Jung was intrigued by the occult, believing that it might well explain some of the mysteries of the human mind. And he was fascinated with the notion that the laws of physics might be used to explain psychic phenomena. Appropriately, he got the idea from discussions with Albert Einstein himself. "It was he," said Jung, "who first started me off thinking about a possible relativity of time as well as space, and their psychic conditionality."

The unconscious mind was like the earth itself, Jung believed, with layer on layer of material containing all of humanity's past experience. Apart from one's personal memories, he held that all human beings shared a collective unconscious containing material common to all humankind. Part of his proof was the endless recurrence of certain symbols in patient dreams and drawings as well as in ancient books and archaeological relics. Jung saw these symbols as representing archetypal human concepts and used them to justify his notion of a universal linkage between people in different dimensions of time. How else, he wondered, could one explain a Hindu graphic symbol called a mandala showing up in the art of the American Navajo?

Hoping to develop a more rigorous exposition of his hunches, Jung took on Nobel laureate and quantum physicist Wolfgang Pauli as his tutor. Pauli shared Jung's view that parapsychology might be a natural bridge between physics and psychology, and he hoped to find a way to express what, in his view, was a natural extension of the quantum phenomena he had helped discover. In 1952, the two scientists collaborated on a book called *The Interpretation of Nature and the Psyche,* each contributing an essay on the subject. The essays put forward the idea of a cosmos, devoid of space and time, in which the psyche and what we normally think of as the universe both exist. Pauli's essay postulated that this cosmos was ordered by its own law, not by human perceptions or the laws of causality. And in this cosmos, mind and matter were not separate, but different, aspects of a single entity.

Jung's essay, which he entitled "Synchronicity: An Acausal Connecting Principle," argued the existence of a timeless unity that incorporated past, present, and future, and where mind and matter were fused as one reality. All his adult life, Jung had been aware of coincident events that were not causally connected but that seemed to defy probability so strongly they could not have happened by mere chance. Knowing that skeptics often use the word *coincidence* to dismiss something others see as paranormal, Jung coined the word *synchronicity* to describe the meaningful coincidence of a psychic and a physical event that have no causal relationship with each other.

is most famous example of synchronicity concerns a patient who was making little progress in her psychoanalysis. One night she dreamed of a golden scarab, an important Egyptian symbol of regeneration. As she described her dream to Jung, an insect flew in the window. The insect proved to be a rose chafer, the closest thing to a golden scarab found in Switzerland. The incident broke down the patient's defenses and led to a new phase in her treatment. For Jung, this was plainly an instance of synchronicity, a meaningful connection.

He found further evidence of synchronicities in the phenomena of lost objects that somehow managed to "find their way back" to their original owners. He cited the case of a German mother who photographed her son in the Black Forest in 1914, just before the start of World War I. The woman took the roll of film to be developed, but the outbreak of fighting made it impossible to collect the pictures.

In his 1919 book The Law of Seriality, Austrian biologist Paul Kammerer said that coincidences did not occur randomly but in series or clusters. He defined seriality as "a recurrence of the same or similar things or events in time or space," calling it "the umbilical cord that connects thought, feeling, science, and art with the womb of the universe which gave birth to them."

Nobel-Prize-winning physicist Wolfgang Pauli described coincidences as "the visible traces of untraceable principles," by which he meant the principles of synchronicity, a theory he helped Carl Jung develop. But the scientific community remembers Pauli best for his work in quantum physics, including being the first to posit the existence of the neutrino, a subatomic particle.

Eventually she realized that she would never see them. In 1916, the same woman visited another shop in a different part of Germany to buy a roll of film to photograph her baby daughter. When the film was developed, every frame turned out to be a double exposure: the new pictures of her daughter on top of the ones taken two years earlier of her son. Somehow her own roll of exposed film had been repackaged as new and then resold to her. For Jung, the episode was a classic example of synchronicity.

As Jung's research drew him more and more into the paranormal, he began to study the ancient arts of astrology and alchemy. He was well aware that he was probing areas that the scientific establishment disdained, but then, he noted wryly, so had Galileo. Jung grew particularly fascinated with the *I Ching,* an ancient Chinese system of divining the future. Although the advice given is generated purely by chance, Jung was astounded by how pertinent the answers to his questions proved to be. After puzzling over the matter for some time, Jung concluded that there must exist a huge network of relationships between people, things, and events and that in some way the *I Ching* helps the human consciousness tune in to this network. The notion seemed to him to be a natural extension of the idea of synchronicity.

Jung freely admitted that not every coincidence had some underlying cosmic significance, but many so-called everyday occurrences, he believed, merited closer scrutiny.

Even seemingly commonplace happenings—such as searching unsuccessfully for a particular book and then finding an abandoned copy of it on a park bench—might well imply a kind of connection in time triggered or powered by the strong significance to the person involved.

One might argue, for example, that the planners of the invasion of Normandy in 1944 must have sent out strong mental signals as they thought and dreamed of D-day. Every phase of the operation was secret and referred to only by code names. And yet, in the month before June 6, the term *overlord,* the secret name for the invasion, and the words *Utah* and *Omaha,* the names for the landing beaches, appeared in crossword puzzles of the *London Daily Telegraph,* along with several other significant code names. Fearing a major breach of security, military police raided the newspaper office. Instead of a Nazi spy, they found a bookish schoolmaster named Leonard Dawe who had been crafting the puzzle for twenty years. Dawe had no knowledge whatever of D-day.

Here and there a particularly odd coincidence is strange enough to make even a skeptic wonder if the earth is not occasionally the butt of some cosmic humor. Consider the murder of Sir Edmundbury Godfrey, reported in the *New York Herald* on November 26, 1911. Three men had savagely beaten their victim to death in a place called Greenberry Hill. The trio convicted and hanged for the crime were named Green, Berry, and Hill.

These sorts of peculiarities were of special interest to a British theorist named John William Dunne, considered by some to have been one of the world's most innovative thinkers concerning the riddle of time. An aeronautical engineer who designed Britain's first military plane, Dunne became interested in the whole subject of time because of what he called the displacement in time represented by his own allegedly precognitive dreams.

One of the dreams that got Dunne started on his life's research happened in 1902, when he was in South Africa during the Boer War. He had a vivid dream, he said, of an island in imminent danger from a volcanic eruption. Dunne saw jets of steam gushing out of fissures in the side of the volcano and knew that if nothing was done to evacuate the area, 4,000 lives would be lost. Not long after, he read grisly newspaper accounts of an eruption of Mt. Pelée on the West Indian island of Martinique reporting that 40,000 had been killed. (Actually, his dream and the reports were both wrong about the number of casualties; some 30,000 were killed.)

A few years earlier, Dunne had begun a lifelong habit of making notes about his dreams shortly after waking each day. His book, *An Experiment with Time,* published in 1927, described his dreams and used them to illustrate his theory of time. Dunne suggested that there are a number of different levels of time and a number of different personalities, or selves, in each of us to observe it. If time "flows," he insisted, then we must have a sense of some other kind of time to measure it by. He called this other time, Time Two. Yet the same limitation must also apply to Time Two, which in turn needs a Time Three from which it can be observed, and so on.

Dunne likened this apparently endless series of time frames to an artist's attempt to paint a picture of the entire universe. Having painted all that he can see, Dunne explained, the artist would realize that something is missing from the picture: himself. To remedy the situation, he must move his easel back and paint himself into the landscape. Still, something would be missing: himself painting himself. Once again, he must move the easel back—and so on. This infinite series of temporal reference points prompted Dunne to give his theory the name *serialism*. If you could trace the series of selves far enough, Dunne wrote, you would come to what he called a "superlative general observer, the fount of all self-consciousness."

One of Dunne's greatest admirers was J. B. Priestley, whose more than 100 books and plays made him one of Britain's most popular authors in the 1930s and 1940s. Priestley shared Dunne's fascination with time, a theme that dominates many of his plays. Although he agreed with many of Dunne's theories, Priestley took issue with the idea of serial time and its infinite series of time frames. He argued that there were only three kinds of time and three selves to experience them—a "self one" that is active when we are awake, a "self two" that observes the world passively when we are asleep or half asleep, and a "self three" that observes everything in a detached fashion. The corresponding times are ordinary time, experienced during such routine activity as waiting for a bus; inner time, experienced during moments of contemplation; and creative time, which is what the name implies and is experienced only rarely.

In 1963, Priestley appeared on television to discuss his ideas, which he was setting forth in a book entitled *Man and Time*. At the close of the program, viewers were invited to write to Priestley about any experiences that seemed to challenge the conventional "common-sense" idea of time. Priestley was deluged with more than a thousand letters. A number of these dealt with what the author called "the influence of the future on the present," cases or dreams that accurately communicate feelings about some future event, although the event itself may not be revealed. One woman described a dream in which, while out walking, she had found herself looking at a strange hospital as inexplicable tears streamed down her cheeks. More than twenty-five years later, she found herself standing outside the same hospital, crying for a friend who had just died there.

Another man wrote that he had suffered in his youth from severe headaches accompanied by visions of brilliant colors. Years later, while serving with the Royal Air Force in Malaya during World War II, the man relived the experience with frightening results. Japanese planes had attacked his truck convoy with bombs and machine-gun fire. A bomb burst close behind him, he wrote, and "the world exploded into a hell of color. All the jagged splinters of red, blue, green and vivid purples caught and swamped me and flung me among the gunners." The man survived the explosion, and the recurring visions of color that had plagued his youth never returned.

Priestley noted that the event was so powerful that it haunted the man for years—and, in a strange way, may have haunted him long before it actually happened. The explosion, Priestley suggested, may have sent its shock waves both forward and backward in time, hitting the man hardest from the future. Following the bomb blast, he remembered the whirl of colors, as one would expect, but as a child he appears to have anticipated them.

Dutch psychoanalyst Joost Meerloo recounts a similar wartime experience, but one in which, he claims, precognition may actually have saved his life. While trying to escape the Nazis during World War II, Meerloo joined a group of people attempting to reach safety in Switzerland. "After fleeing from Holland to Belgium," he later recalled, "I realized that our so-called secret escape had been poorly organized. We took refuge in a brothel in Antwerp, where we could hear the German soldiers singing their heads off. We were relatively safe, but I was convinced we could be treacherously delivered into the hands of the enemy. That

confidence and laughed at my nightmarish fears.''

The next day, Meerloo's group was indeed captured while riding a train to Paris. His dream had now taken on an urgent significance, and he believed that his life depended on remaining apart from the other prisoners.

''We spent the night on the Belgian-French border and after a while the French police brought a motley crew of Belgian money smugglers into the prison where we all sat and waited. The police asked the SS guards to watch over the smugglers until they were called for. We waited out the long night, full of tension and anxiety with armed guards all around us. Instinctively I moved over to the side where the smugglers were huddled and mingled with them. When the French police came for them in the early morning, the German guards, thinking I belonged with these criminals, let me go too.''

Meerloo's actions, guided by his dream vision, had saved his life. ''In the French prison I told the guards my story and after some delay they let me go. The design of my dream had come true. Nothing was ever heard again of the other members of the original group.''

For Meerloo, the terrifying experience offered compelling evidence of precognition. ''Our thinking,'' he wrote, ''always rushes back and forth in time. The future goal we want to reach directs our course right now, while at the same time unconscious ancient patterns are acted out. This dual temporal determination of both conscious and unconscious intentions points to a complicated relationship. The slogan 'The

night, before falling off to sleep, I resolved to go it alone and separate myself from the other escapees. Then I dreamed I was caught and imprisoned with the others. But in my dream I kept to myself. We were herded together with other prisoners and I was making plans to escape with that group. I awoke with a feeling of self-end justifies the means' overlooks the fact that the very means which are used will always contaminate the end result. Thus the end will *not* justify the means; influences run *both* ways in time.''

Both Meerloo and Priestley appear to agree that the mind is influenced by impulses from the past as well as the future. Reported precognitions like the one that is supposed to have helped save Meerloo's life are clearly the most dramatic examples of what Priestley called the influence of the future on the present. The notion of retrocognition, the knowledge of past lives or events, is—if true—an equally dramatic illustration of the influence of the past. Stories of retrocognitive experiences are often offered as evidence of reincarnation, but another theory suggests that the receiver of information from the past is somehow able to scan across the time barrier and view the past lives of others.

British television viewers got an eerie demonstration of seeming retrocognition when hypnotist Arnall Bloxham placed a subject into a deep trance and caused her to recall a past life. Under hypnosis, the subject told a frightening story of a woman named Rebecca the Jewess, who lived in York, England, in 1189—a time of violent religious intolerance. ''Rebecca'' described how she had hidden from her persecutors in the crypt of Saint Mary's Church, Castlegate, York.

For a while, Rebecca's refuge was secure, but she was discovered and killed during a massacre of Jews the following year. Rebecca's detailed description of her death was shockingly painful and graphic, including details that a present-day person supposedly could not have known.

Saint Mary's, the church in Castlegate where Rebecca claimed to have hidden, was still standing 800 years later. A TV camera crew visited the site in the hope of verifying the strange tale but found no crypt. Six months later, however, workers digging beneath the chancel of Saint Mary's uncovered the secret hiding place—just as Rebecca had described it.

Believers say such cases present a fascinating opportunity to study the ways in which the past can exert itself upon the present. An even more dramatic phenomenon is the so-called time slip. As in the case of Buterbaugh, the college secretary who apparently believed that she briefly moved into an earlier time, people who claim to have experienced time slips say they not only saw into another time period but actually felt the sensation of being there. A time slip, they say, is usually sudden and is sometimes accompanied by a tin-

Author J. B. Priestley contemplates letters from television viewers about strange time experiences. Approximately three-quarters of those who reported premonitions or dreams that eventually came true were women—indicating, he said, that men ''are more likely to inhibit themselves for the sake of appearing sensible.''

gling of the skin or a feeling of nausea.

One of the strangest reports of a time slip in recent years came from a village near Lake Neuchâtel in Switzerland in the early 1960s. A British tourist—identified in accounts of the case as Mr. B—was with two American companions, Charles Muses, a researcher and writer on parapsychological subjects, and Muses's wife. Visiting the ruins of an old Roman theater, the three found that much of the stone stage remained, as well as some of the tiers of seats that rose up the sides of the theater with high walls behind them. Mounting the stage, Mr. B decided to test the theater's acoustics by standing center stage

Aeronautical engineer and time theorist J. W. Dunne, seen here in a biplane he designed, called his theory of serial time and multiple selves "the first scientific argument for human immortality."

and speaking in a normal voice. His friends called back from the top row of seats that they could hear him perfectly. Then he stepped down into the orchestra pit.

Suddenly, he said, he seemed to slip centuries into the past. According to his story, the crumbled sides of the theater became whole, and missing rows of seats appeared as if from thin air. As he stood gaping in shock, a large crowd of people in Roman dress poured in and began filling the seats. Scanning the unfamiliar scene, he noticed a very pretty girl dressed in yellow in one of the upper rows. As he

watched, she stared straight at him and then plucked at the clothing of her companions and pointed him out with her finger, apparently shocked by his appearance. Her friends seemed unable to see him. A moment later a young man walked on stage and began playing a lute of ancient form. Mr. B said he heard only the first notes; almost at once, the music began to fade away. As it did, the vision of the Roman theater and the girl in yellow vanished, and he found himself once again in the present. He later said he had found the experience enjoyable and regretted that it ended so quickly. Relating the experience to a writer, he exclaimed, "How strange I must have looked to a Roman girl!"

The vast majority of time-slip stories involve the past, but occasionally there is a report of a slip into the future. According to one such account, a Miss R. H. Hodgskin of Birmingham, England, and a friend identified as Tessa G. visited the Tower of London on April 20, 1974. There they spent some time in the armory looking at the ancient guns, swords, and axes. The women said that they felt oppressed by the place and were climbing the stairs to leave when Tessa G. asked her friend if she heard children crying. No, said Miss Hodgskin, the only thing she could hear was the voices of other tourists.

But by that time, Tessa appeared extremely worried. "I can hear children shouting and screaming," she insisted.

Betting on a Dream

It was a dream that could be called a bookie's nightmare. According to psychical researcher H. E. Saltmarsh, on May 31, 1933, a man named John H. Williams dreamed that he heard a radio broadcast of the Derby, England's most prestigious horse race—scheduled to be run that very afternoon. Williams told friends of his dream, including the names of two of the first four finishers: Hyperion and King Salmon.

Although Williams was a Quaker who opposed gambling, curiosity drove him to listen to the race on the radio. He was reportedly astonished to hear the announcer repeating the commentary he had dreamed—and even more astonished to hear the name of the winner: Hyperion. King Salmon ran second. Investigator Saltmarsh said he confirmed the tale with Williams's friends. Whether they had bet on his dream horses was not recorded.

Skeptics note that, given the countless number of dreams about the outcome of races or other forms of gambling, some inevitably will prove correct. But paranormal researchers contend there are many documented accounts of accurate precognitive dreams of soccer matches, lotteries, and other betting events.

One of the most famous of these again involved horse racing. On March 8, 1946, an Oxford undergraduate named John Godley (later Lord Kilbracken) dreamed that he read the racing results and noted that two horses, Bindal and Juladin, had both won their races at 7-1 odds. When he awoke he checked the paper, saw that two such horses were scheduled to race that day, and placed healthy bets on each. Both won. Off and on over the next twelve years, Godley dreamed of race results and garnered both fame and fortune as his predictions—many of which were documented—paid off. Then, as abruptly as it came, his gift for seeing the future disappeared.

Still Miss Hodgskin heard nothing unusual. According to the report, both women forgot about the incident until some months later, when a terrorist bomb exploded in the armory of the Tower of London, killing and wounding several people, including some children.

The apparently random nature of reported time-slip experiences has baffled many researchers. Many students of parapsychology say that if science could uncover the causes of the phenomena, we would at last understand the elusive link between the secrets of the mind and the mystery of time. British archaeologist Thomas Lethbridge, once Keeper of Anglo-Saxon Antiquities at Cambridge University, spent much of his life probing this puzzle.

Lethbridge said his interest in the paranormal began when he discovered that he possessed a talent for dowsing, the ability to locate water or other buried objects with a divining rod or pendulum. He began dowsing in the early 1930s, while searching for Viking graves on Lundy Island in the Bristol Channel. The island was honeycombed with volcanic rock seams hidden under layers of slate, and Lethbridge decided to try to find them using a forked hazel twig as a dowsing rod. With a blindfold fastened over his eyes, he lurched back and forth across the island. Lethbridge said that each time he passed over a deposit of volcanic rock, the twig in his hands twisted downward.

He decided that the volcanic rocks must radiate a faint magnetic energy that he was somehow able to pick up. Based on his experiments over the years, he speculated that certain areas of the earth have individual electric fields that may be able to record human emotions, thoughts, and even images. These energies might later manifest themselves to other people as feelings or apparitions. Lethbridge himself claimed to have experienced this phenomenon. One day, while walking through a forest with his mother, he started feeling very depressed when they reached a certain spot. A few days later, he learned that the body of a suicide had been discovered there.

As Lethbridge developed what he called his theory of natural "tape recording," he put aside the dowsing rod in favor of the pendulum, an ancient device used for such tasks as determining the sex of an unborn baby or finding buried metal. Lethbridge claimed his experiments showed that a pendulum reacts consistently to many different substances, the device's response to a particular material depending on the length of the string. For example, he said, a pendulum with a seven-inch string began swinging when held near sulphur; a twenty-nine-inch pendulum reacted to gold and a fourteen-inch to glass.

He was even more excited by his next alleged discovery. The pendulum, Lethbridge said, appeared to react not only to objects but also to human thought. According to this astonishing assertion, if he concentrated on the moon, the pendulum reacted in a certain way. And it reacted differently to thoughts of each of the four points of the compass.

Lethbridge never claimed that the pendulum itself had paranormal power. Instead he saw the pendulum as a way of tapping areas of the human mind that hold mysterious knowledge and power but that cannot convey them directly to the conscious mind. The pendulum, he believed, unlocked this data by allowing the unconscious mind to move the body's muscles in such a way that the pendulum seemed to supply the answers out of thin air.

Since he thought that both dowsing rods and pendulums reacted to energy or magnetism from the earth, Lethbridge guessed they would respond even more powerfully in places where natural magnetism is stronger. He apparently believed that Britain's megalithic sites, such as Stonehenge and Glastonbury, would demonstrate greater-than-normal amounts of natural magnetism. In particular, he became intrigued with the so-called ley lines first discovered by a brewer named Alfred Watkins in the early 1920s. Watkins had noticed that old footpaths and farm tracks in rural England seemed to form a network of lines in the earth, connecting old churches, ancient mounds, standing stones, cairns, and other supposedly sacred sites. Watkins said he found that he could uncover lost trackways simply by connecting the ancient sites with straight lines traced on a map. He reported that when he traveled into the country to search out the trackways, he found huge sections of the original lines exactly as he had drawn them on his map.

atkins's work fascinated Lethbridge. Using a pendulum and dowsing rod, Lethbridge quickly decided that the earth's magnetic forces were strongest where the leys crossed. He had read in other researchers' reports that time slips and other psychic phenomena tend to recur at specific locations. Lethbridge suggested that ley crossings might well act as some sort of trapdoor between parallel dimensions, allowing us to move from one world to another. These crossings, he believed, might one day reveal the secret to any number of mysterious human disappearances and perhaps unlock the riddle of UFOs, which he said are seen more frequently in areas where several leys converge.

Some parapsychologists say that if these places actually do represent some sort of doorway to different dimensions, as Lethbridge suggested, then they may account for at least a few of the disappearances that over the centuries have seemed to defy explanation. One such case is that of Benjamin Bathurst, British ambassador to the court of Francis I of Austria. On November 25, 1809, the ambassador stopped at an inn in Perlberg on his way back to London. Accompanied by his valet and secretary, Bathurst had paused for something to eat. After a short rest, the ambassador strolled out into the courtyard and prepared to climb into his carriage. He told his valet to wait at the carriage door while he inspected the horses. According to one historical account, "He stepped round to the heads of the horses—and was never seen again." There were no clues or signs of foul play, or so the story says. The valet heard and saw nothing. Despite a long and intensive investigation by his family and the British government, the ambassador was never found. Tales of supposedly baffling disappearances

Strange feelings of fear, depression, and suicide enveloped Professor Thomas Lethbridge and his wife, Mina, on a particular cliff at Latham Bay in Devon, England (above). Lethbridge, a Cambridge ethnologist and archaeologist, suggested that the magnetic fields of water and other natural elements may "record" human emotions, thoughts, and images from the past—or even the future. Nine years after the Lethbridges' experience, a man, perhaps unable to resist the same urgings, committed suicide by leaping from the cliff.

abound. They include the case of a prominent editor of *Harper's Weekly* who disappeared in New York City and was never found and that of wealthy socialite Dorothy Arnold, who went shopping on Fifth Avenue in 1910 and seemed to vanish into thin air.

But some much-circulated disappearance stories do not stand up to close scrutiny. One famous example involves a whole World War I British regiment, the First Fourth Norfolk, numbering more than a thousand men. One day in August of 1915, says the story, the regiment launched an attack against the Turks in Gallipoli, disappeared into a cloud bank, and remains missing to this day. But suggestions that the First Fourth Norfolk was wrenched into another dimension or another time begin to sound questionable when factual accounts of the event are considered. The advance against the Turks culminated in a truly terrible battle, so ferociously bloody that the crews who arrived later to collect the dead could not find enough whole bodies to account for all the missing men—although there was grisly fragmentary evidence aplenty that many more had died there. Under the circumstances, it would not have been surprising if the British military authorities themselves had encouraged the story of mysteriously vanishing soldiers.

Another well-worn—but in the final analysis, flimsy—account is that of the supposedly inexplicable vanishing of one David Lang, a Tennessee farmer who is said to have stepped off the face of the earth in September of 1880. The story, which is replete with names, dates, and other convincing-sounding details, begins with Judge August Peck and a companion arriving for a visit to the Lang farm near Gallatin, Tennessee. As Peck, a friend of the Lang family, pulled up to the property in his horse and buggy, Mrs. Lang and children George and Sarah rushed from the farmhouse to greet him. Beyond them, Judge Peck saw David Lang walking across his field. The farmer waved as he advanced toward the group a few hundred yards away.

According to the oft-published story, David Lang—without warning, and in plain sight of the judge, his friend, and the rest of the Lang family—seemed simply to slip from

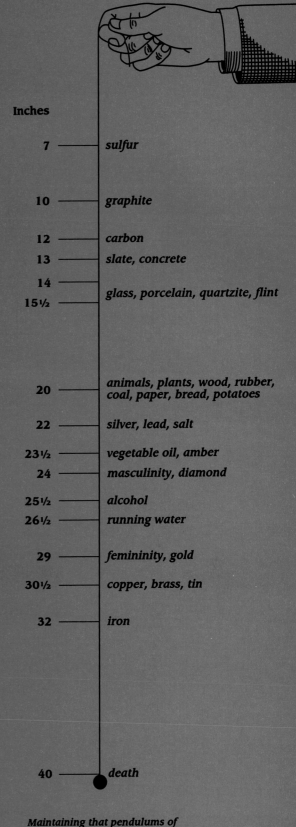

Inches

7 —— *sulfur*

10 —— *graphite*

12 —— *carbon*

13 —— *slate, concrete*

14 ——

15½ —— *glass, porcelain, quartzite, flint*

20 —— *animals, plants, wood, rubber, coal, paper, bread, potatoes*

22 —— *silver, lead, salt*

23½ —— *vegetable oil, amber*

24 —— *masculinity, diamond*

25½ —— *alcohol*

26½ —— *running water*

29 —— *femininity, gold*

30½ —— *copper, brass, tin*

32 —— *iron*

40 —— *death*

Maintaining that pendulums of different lengths swing in response to particular materials or thoughts, Lethbridge laid out a table of pendulum lengths for divining various substances and qualities. He said that pendulums of the right length would also respond to thoughts of youth, anger, danger, and life.

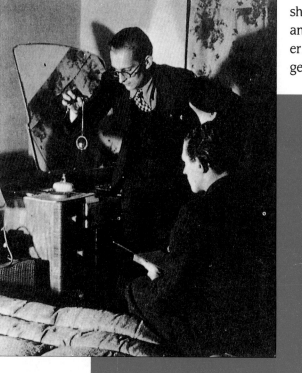

view. All five witnesses raced to the spot, expecting to discover that Lang had fallen into some sort of hole. Instead they found an unbroken stretch of earth baked hard by the summer sun. As Mrs. Lang began sobbing hysterically, Judge Peck summoned the Lang's neighbors and began a systematic search of the entire forty-acre meadow where the farmer had vanished. By nightfall, the searchers had found no sign of the missing man.

Present-day accounts say that David Lang's disappearance was widely reported in the local newspapers and that as the tale spread, curiosity seekers began to arrive at the farm to view the spot where he had vanished. The search for the missing farmer persisted for more than a year. Authorities even dug into the soil at the spot where he had last been seen in the hope of uncovering some sort of underground cavern. They found only solid limestone.

The following year, says the story, the two Lang children, Sarah and George, noticed that the remaining grass at the spot had grown brown and coarse and that their cattle would not graze there. Sarah, who was then twelve years old, became very excited, believing that she sensed her father's presence in the area. She called out aloud to him but received no reply. Then, as the children walked away, they heard a man's voice shouting for help.

At once the children brought their mother to the spot. Mrs. Lang also called out to her husband and heard a thin voice answer as though from very far away. Immediately the neighbors organized a new search for Lang, but to no avail. Each day, George and Sarah returned to the spot and shouted to their father. Each day, his answering cry grew fainter and fainter. After a time, it faded away altogether. David Lang was never heard from again. One possible explanation frequently offered in published accounts is that perhaps the farmer stepped through some kind of rip in the space-time continuum into another dimension.

Recently, a pair of British investigators attempted to verify the details of the Lang vanishing. The results of their inquiry cast grave doubt on the tale's authenticity. They could not even confirm that anyone named David Lang, or August Peck, had lived anywhere near Gallatin at the time in question. Local papers of the period reveal no stories of any such disappearance. The pair concluded that the entire story may well have been concocted by an imaginative traveling hardware salesman named Joe Mulhatten, who was known to entertain his family and friends with strange tales supposedly gathered on his travels. In fact, the salesman's flights of fancy were so well known that his listeners came to refer to his tall tales as Mulhattens.

Not all mysterious disappearances can be so easily dismissed as "Mulhattens," however. The Bermuda Triangle, an area of ocean between Bermuda, Puerto Rico, and the coast of Florida has been the focus of mystery and superstition since the days of Columbus. By some tallies, at least 147 ships, boats, and aircraft remain unaccounted for in modern times. The most famous case is the disappearance of Flight 19—five U.S. Navy planes lost in December 1945 and never seen again. Messages from the pilots stated that their gyrocompasses were "going crazy," and they were unable to determine their position. Students of the occult blame the alleged mysterious powers of the Bermuda Triangle, which are often associated with doorways to other dimensions; skeptics attribute Flight 19's disappearance to the normal hazards of aviation over water. It should be noted, however, that many other reports of Bermuda Triangle vanishings include garbled radio messages describing the erratic behavior of equipment and sudden loss of power.

The puzzle of areas like the Bermuda Triangle was of particular interest to Charles Hoy Fort, an obscure newspaperman from Albany, New York, who spent twenty-six years of his life collecting reports of unexplained events. Fort was obsessed with every manner of unusual happening—the appearance of people from nowhere, spontaneous human combustion, unidentified flying objects, lights on the moon, stigmatic wounds, rains of stones and blood, wolf children and wild men, teleportations, visions, and levitations. Years of combing the principal newspapers and scientific journals of the world nearly blinded him, but Fort managed to publish four books of what he called the damned—weird, unexplainable events that the scientific community preferred to ignore.

Fort rarely had any explanation for the odd events he collected. It was enough for him to state the facts as baldly as possible and then acidly quote from unconvincing official explanations. Still, surveying his collection of tens of thousands of events, written on small squares of paper in a cramped private shorthand, certain beliefs emerge. He regularly expressed doubts about the theory of evolution, thought

that psychic power could one day become a weapon of war, and believed that the earth had been visited—and possibly fought over—by ancient astronauts from outer space.

Undoubtedly an eccentric, Fort was also a prodigious researcher. Of more than 65,000 paranormal events he documented, only 1,200 or so of those he deemed authentic are mentioned in his books. A cynic

British author Colin Wilson found that a whalebone dowsing rod "jumped upwards" at this stone circle, called the Merry Maidens, near Penzance, England. He says that physicists using gaussmeters, which measure electromagnetism, discovered a powerful magnetic field in one such standing stone.

about scientific explanations, he believed scientists too often let their own prejudices color their theories and was outraged when data not fitting the accepted views was ignored, discredited, or explained away. Some psychical researchers now regard him as a visionary, and writers of fantasy and science fiction have seized on his ideas, many of which have become part of the intellectual currency of our time.

Appearances and disappearances held a special fascination for Fort. He attributed many mysterious appearances and vanishings to a natural force he called teleportation—an earthly power capable of moving objects. Fort speculated that this force had once been enormously potent—strong enough to help shape the earth and perhaps scatter life forms among various planets. In its modern, vestigial form, he believed, teleportation accounted only for minor phenomena, such as rains of stones or fish.

Fort was also among the first to give serious consideration to the existence of unidentified flying objects. He reasoned that if aliens had appeared on earth in earlier times they would almost certainly have been mistaken for apparitions or demons. Many historical accounts of ghosts and goblins, he suspected, might actually be chronicles of visitors from other worlds.

This view is now widely shared among paranormal investigators, and Fort's research forms the foundation of several contemporary theories linking reported UFO phenomena to space-time anomalies. A number of UFO researchers assert that unidentified flying objects may well represent some form of breach of the time barrier. The rapid movements and sudden appearances and disappearances

The 1932 New York Times obituary of Charles H. Fort called him a foe of science. More recent evaluations portray Fort as a serious researcher, the first to make systematic studies of such phenomena as fish falls (bottom right), flaming planks that fell from the sky over Touraine, France, in 1670 (top left), and other strange "rains" (top right). He was impatient with scientists who ignored evidence of events that they could not explain.

Intriguing Speculations on a Flat Idea

Assuming that UFOs exist, is it possible that they come not from outer space but from another dimension, one that our experiences in a world of three spatial dimensions keep us from understanding? The principles behind that question—but not the specific question itself—were pondered more than a century ago by a Victorian schoolmaster named Edwin A. Abbott.

In his novel, *Flatland: A Romance of Many Dimensions,* Abbott explored the disturbing effects of being suddenly exposed to an unperceived additional dimension. His book, published in 1884, presented ideas that can still inspire far-ranging speculation about several unexplained phenomena, including not only UFOs but sudden unaccountable appearances and vanishings and so-called hauntings.

A satirical story that mocks the rigid and often heartless Victorian society in which Abbott lived, *Flatland* describes a rather smug two-dimensional world whose citizens are flat geometric shapes sliding about on a flat surface, somewhat like coins on a tabletop. They have no words to describe concepts such as "up" or "down" and can move only forward, backward, and sideways on the same plane.

One day, a three-dimensional Sphere visits Flatland. He makes contact with a Flatlander named A Square and exposes him to the concept of a three-dimensional world. When A Square tries to teach his fellow Flatlanders about the third dimension, he is locked up as a threat to society.

The three-dimensional Sphere natu-rally would baffle and terrify the two-dimensional Flatlanders, since they could see only two of his dimensions at once—a flat slice through his body that grew larger or smaller as the Sphere moved up or down. Flatlanders, with no knowledge of the vertical dimension, would see a creature rapidly changing its shape and then disappearing once it moved above or below their plane. Any Flatlander who had not seen the Sphere personally would be reluctant to believe it existed.

The ideas Abbott presented have been the stuff of science fiction and speculative discussion about the paranormal ever since. What if there are more dimensions than the four we are used to—time and three spatial dimensions? (It is not an unwarranted question: Some physicists seriously suggest there may be as many as twenty-six dimensions in all.) If a being or object that normally existed in other dimensions were to pass through ours, might not our view of it be as confusing as the Flatlanders' view of the Sphere?

On this basis, it has been suggested that UFOs might come from some other-dimensional universe to pop in and out of ours. If their journeys only briefly intersect our dimensions, ask believers in this theory, wouldn't they appear and disappear with remarkable speed—possibly collecting human beings to take with them, as so many self-proclaimed "abductees" allege? Could accidental encounters with random holes into other dimensions account for people or things occasionally seeming to vanish into thin air? For unusual "rains" of objects from clear skies? For what we call ghosts?

The prospects are disturbing as well as fascinating. Flatland's A Square found a trip to the third dimension unsettling: "An unspeakable horror seized me. There was a darkness; then a dizzy, sickening sensation of sight that was not like seeing; I saw a Line that was no Line; Space that was not Space: I was myself and not myself. When I could find voice, I shrieked aloud in agony, 'Either this is madness or it is Hell.' 'It is neither,' calmly replied the voice of the Sphere, 'it is Knowledge; it is Three Dimensions: open your eye once again and try to look steadily.' "

Edwin A. Abbott

attributed to UFOs, if true, would seem well beyond the capability of any spacecraft our technology could conceive. Even a craft capable of traveling at the speed of light would take years to reach earth from the nearest star. Such physical realities have pushed some theorists toward paranormal explanations, such as the existence of parallel universes, trapdoors in space, and even the ability to manipulate time.

Writer Guy Lyon Playfair thinks UFOs may offer proof of a universe with four spatial dimensions. Just as particle physicists deal in matter that is largely invisible, Playfair posits the existence of a fourth dimension composed of spirit matter that is able to move through our universe undetected. This spirit matter, he postulates, may exist in some form of hyperspace, with an "intelligent source" capable of moving objects in and out of its dimension. These speculations suggest a plausible-sounding explanation for the reports of seemingly impossible high-speed maneuvers and sudden appearances and disappearances of UFOs: They could be slipping in and out of a different dimension.

Another writer, Donald A. White, offers a different ex-

A dog-eared science-fiction magazine suggests that UFOs— which some people speculate come from another time or an other-dimensional universe— may have long been responsible for strange vanishings, in this case the supposedly mysterious disappearance of ancient Cambodians from Angkor in 1431.

SPACE SHIPS AT ANGKOR WAT?
Was it really a fleet of space ships that came to Angkor Wat and removed its population to another planet? Anyway, its people could not have vanished more completely than if this had been a reality!

planation for the apparently impossible speed and movement attributed to UFOs. Many people who claim to have seen UFOs say the objects appear to pulsate with color, often shifting from yellow to red or green to blue, as they become visible to the naked eye. White believes that this phenomenon is a cosmic variation of the Doppler effect, with the colors changing up and down the spectrum as the speed of the UFO increases or decreases. It is, says White, what you would most likely see if a craft were traveling close to the speed of light—or if it were crashing through the barrier of time; people may spot UFOs only when they slow down.

Do UFOs, if they actually exist, hold the answer to the mystery of time? It is ironic to think that humankind may have to solve one riddle to get at the other. Yet it is perhaps stranger still to think, as some paranormal researchers believe likely, that the secret of time may lie in such everyday happenings as odd coincidences and precognitive dreams. A sensation of familiarity upon entering an unfamiliar room. A book that flips open to a significant passage. A dream of a long-lost friend who then telephones unexpectedly. Some believe that these and a thousand other commonplaces may yield up a tremendous secret—in time.

The writer Joost Meerloo, whose life was supposedly saved by a strange and prophetic dream, once had the chance to discuss meaningful coincidences with Albert Einstein himself. Meerloo says that the physicist was at first highly skeptical that any practical results could ever be obtained in the study of coincidence.

Meerloo paused, undoubtedly mulling over how the insights of this great thinker had advanced humankind's understanding of time and space by almost incredible lengths. Then he persisted: "But are you yourself not such a happy coincidence?"

Einstein paused for a moment and then smiled. "This argument you win," he said.

ACKNOWLEDGMENTS

The editors thank these individuals and institutions for their valuable assistance in the preparation of this volume: Dr. Halton Arp, Max Planck Institut für Astrophysik, Garching, West Germany; François Avril, Conservateur, Département des Manuscrits, Bibliothèque Nationale, Paris; Professor Hanno Beck, Bonn, West Germany; Conte Guglielmo Pennati Beluschi, Monza, Italy; Dr. Herbert Bibach, Max Planck Institut für Verhaltensphysiologie, Andechs, West Germany; Contessa Maria Fede Caproni, Rome; Augusto Foà, Dipartimento di Scienze del Comportamento Animale, Università di Pisa, Rome; Andrea Galvagno, Pioneer Explorations and Researches, Ancona, Italy; Guido Galvagno, Pioneer Explorations and Researches, Ancona, Italy; Professor Ernst Heldmeyer, Zoologisches Institut, Universität Marburg, West Germany; Beate Janouschek, Zoologisches Institut, Universität Frankfurt, West Germany; Edouard Lambelet, Cairo; Dorothy Mastricola, The Time Museum, Rockford, Illinois; Augusto Meratti, Monza, Italy; Maurizio Montalbini, Pioneer Explorations and Researches, Ancona, Italy; Floriano Papi, Dipartimento di Scienze del Comportamento Animale, Università di Pisa, Rome; Dr. Dorie Reents-Budet, Curator of Pre-Columbian Art, Duke University Museum of Art, Durham, North Carolina; Professor Alain Reinberg, Directeur de Recherches, CNRS, Paris; John Ross, Cortona, Italy; Ann Stevens, London; Professeur Yvan Touitou, Biologiste des Hôpitaux Faculté de Médecine Pitié Salpêtrière, Paris.

BIBLIOGRAPHY

Abbott, Edwin A., *Flatland*. New York: Dover, 1953.

Baker, Robin, ed., *The Mystery of Migration*. New York: Viking Press, 1981.

Bancroft, Anne, *Zen: Direct Pointing to Reality*. London: Thames and Hudson, 1979.

Barnet, Sylvan, *Zen Ink Paintings*. Tokyo: Kodansha International, 1982.

Barnett, Lincoln, *The Universe and Dr. Einstein*. New York: Time, 1962.

Bartusiak, Marcia, "Before the Big Bang: The Big Foam." *Discover*, September 1987.

Benford, Gregory, "Alan Guth, Cosmological Physicist." *Omni*, November 1988.

Blacker, Carmen, and Michael Loewe, eds., *Ancient Cosmologies*. London: George Allen & Unwin, 1975.

Boorstin, Daniel J., *The Discoverers*. New York: Random House, 1983.

Brod, Craig, *Technostress: The Human Cost of the Computer Revolution*. Reading, Mass.: Addison-Wesley, 1984.

Brunés, Tons, *The Secrets of Ancient Geometry and Its Use*. Transl. by Charles M. Napier. Vol. 2. Copenhagen, Denmark: Rhodos, 1967.

"Bubbles upon the River of Time." *Science*, February 1982.

Burland, C. A., and Werner Forman, *Feathered Serpent and Smoking Mirror*. New York: G. P. Putnam's Sons, 1976.

Burlingame, Roger, *Dictator Clock: 5,000 Years of Telling Time*. New York: Macmillan, 1966.

Caes, Charles J., *Cosmology*. Blue Ridge Summit, Pa.: Tab Books, 1986.

Calder, Nigel, *Einstein's Universe*. New York: Penguin Books, 1980.

Campbell, Jeremy, *Winston Churchill's Afternoon Nap*. New York: Simon and Schuster, 1986.

Campbell, Joseph:
The Masks of the Gods. New York: Viking Press, 1972.
The Mythic Image. Princeton, N.J.: Princeton University Press, 1974.
Mythologies of the Great Hunt. Vol. 1, Part 2 of *Historical Atlas of World Mythology*. New York: Harper & Row, 1988.

Capra, Fritjof, *The Tao of Physics*. Boston: Shambhala, 1985.

Cavendish, Richard, *The Great Religions*. New York: Arco, 1980.

Cavendish, Richard, ed.:
Man, Myth & Magic. New York: Marshall Cavendish, 1985.
Mythology: An Illustrated Encyclopedia. London: Orbis, 1980.

Chaisson, Eric, *Relatively Speaking: Relativity, Black Holes, and the Fate of the Universe*. New York: W. W. Norton, 1988.

Cheney, Sheldon, *Men Who Have Walked with God*. New York: Alfred A. Knopf, 1948.

Ching, Francis D. K., *Architecture: Form, Space & Order*. New York: Van Nostrand Reinhold, 1979.

Chorlton, Windsor, and the Editors of Time-Life Books, *Cloud-Dwellers of the Himalayas: The Bhotia* (Peoples of the Wild series). Amsterdam: Time-Life Books, 1982.

Clark, Ronald W., *Einstein: The Life and Times*. New York: Avon, 1979.

Close, Frank, Michael Marten, and Christine Sutton, *The Particle Explosion*. New York: Oxford University Press, 1987.

Cook, Roger, *The Tree of Life: Image for the Cosmos*. New York: Avon, 1974.

"The Crash of '87 Sends Stock Markets around the World Plummeting to Record Losses." *Time*, November 2, 1987.

Croft, Peter, *All Color Book of Roman Mythology*. London: Octopus Books, 1974.

Darling, David, "The Quest for Black Holes." *Astronomy*, July 1983.

Davidson, H. R. Ellis, *Gods and Myths of Northern Europe*. New York: Penguin Books, 1982.

Doczi, György, *The Power of Limits: Proportional Harmonies in Nature, Art and Architecture*. Boston: Shambhala, 1981.

Doggett, Rachel, ed., with Susan Jaskot and Robert Rand, *Time: The Greatest Innovator* (exhibition catalog). Washington, D.C.: The Folger Shakespeare Library, 1986.

Eliade, Mircea, *The Sacred and the Profane*. San Diego: Harcourt Brace, 1957.

The Encyclopedia of Eastern Philosophy and Religion. Boston: Shambhala, 1989.

Evans, Joan, ed., *The Flowering of the Middle Ages*. New York: Bonanza Books, 1985.

Ferris, Timothy, *Coming of Age in the Milky Way*. New York: Doubleday, 1988.

Fish & Wildlife Service, *Migration of Birds*. Circular 16. Washington, D.C.: United States Department of the Interior, 1979.

Fisher, Allan C., Jr., "Mysteries of Bird Migration." *National Geographic*, August 1979.

Fisher, Leonard Everett, *Calendar Art: Thirteen Days, Weeks, Months, and Years from around the World*. New York: Four Winds Press, 1987.

Flacelière, Robert, *Greek Oracles*. Transl. by Douglas Garman. New York: W. W. Norton, 1965.

Flanagan, Dennis, *Flanagan's Version: A Spectator's Guide to Science on the Eve of the 21st Century*. New York: Random House, 1989.

Fleet, Simon, *Clocks*. New York: G. P. Putnam's Sons, 1961.

Forward, Robert L., *Future Magic*. New York: Avon, 1988.

von Franz, Marie-Louise, *Time: Rhythm and Repose*. London: Thames and Hudson, 1978.

Fraser, J. T.:
Of Time, Passion, and Knowledge. New York: George Braziller, 1975.
Time: The Familiar Stranger. Amherst, Mass.: The University of Massachusetts Press, 1987.

Fraser, J. T., ed., *The Voices of Time*. New York: George Braziller, 1966.

Freedman, David H., "Cosmic Time Travel." *Discover*, June 1989.

Freund, Philip, *Myths of Creation*. New York: Washington Square Press, 1965.

Friedman, Milton, "Chungliang Al Huang: A Master of Moving Meditation." *New Realities*, May/June 1989.

Gardner, Martin, "Can Time Go Backward?" *Scientific American*, January 1967.

Gaskell, G. A., *Dictionary of Scripture and Myth*. New York: Dorset Press, no date.

Gebauer, Paul, *Spider Divination in the Cameroons*. Milwaukee: Milwaukee Public Museum, 1964.

Gillispie, Charles Coulston, ed., *Dictionary of Scientific Biography*. Vol. 5. New York: Charles Scribner's Sons, 1981.

Goldberg, Joshua, *Tibetan Tankas* (exhibition catalog). Tucson, Ariz.: The University of Arizona Museum of Art, 1980.

Gorman, Peter, *Pythagoras: A Life*. London: Routledge & Kegan Paul, 1979.

Goswamy, B. N., *Essence of Indian Art* (exhibition catalog). San Francisco: Asian Art Museum of San Francisco, 1986.

Gott, J. Richard, III, "Creation of Open Universes from de Sitter Space." *Nature*, January 28, 1982.

Goudsmit, Samuel A., Robert Claiborne, and the Editors of Time-Life Books, *Time* (Life Science Library). Alexandria, Va.: Time-Life Books, 1980.

Grant, Campbell, *Canyon de Chelly: Its People and Rock Art*. Tucson, Ariz.: The University of Arizona Press, 1978.

Great Religions of the World. Washington, D.C.: National Geographic Society, 1971.

Gribbin, John, *Spacewarps*. New York: Dell, 1983.

Guthrie, Kenneth Sylvan, comp. and transl., with additional translations by Thomas Taylor and Arthur Fairbanks, Jr., *The Pythagorean Sourcebook and Library*. Ed. by David R. Fideler. Grand Rapids: Phanes Press, 1987.

von Hagen, Victor Wolfgang, *The Ancient Sun Kingdom of the Americas*. Cleveland: World, 1961.

Hall, Angus, *Signs of Things to Come*. Garden City, N.Y.: Doubleday, 1975.

Hall, Edward T.:
The Dance of Life: The Other Dimension of Time. Garden City, N.Y.: Doubleday, 1983.
The Hidden Dimension. Garden City, N.Y.: Doubleday, 1966.

Halle, Louis J., *Out of Chaos*. Boston: Houghton Mifflin, 1977.

Hammond, Norman, *Ancient Maya Civilization*. New Brunswick, N.J.: Rutgers University Press, 1982.

Harris, Melvin, "Once in a Lifetime." *The Unexplained* (London), Vol. 12, Issue 136.

Harrison, Edward, *Masks of the Universe.* New York: Macmillan, 1985.

Harthan, John, *The Book of Hours.* New York: Park Lane, 1982.

Hawking, Stephen W., *A Brief History of Time.* New York: Bantam Books, 1988.

Heath, Thomas, *A History of Greek Mathematics.* Vol. 1. Oxford: Clarendon Press, 1921.

Heggie, Douglas C., *Megalithic Science: Ancient Mathematics and Astronomy in North-West Europe.* London: Thames and Hudson, 1981.

"Helping Workers Stay Awake at the Switch." *Business Week,* December 8, 1986.

Herbert, Wally, and the Editors of Time-Life Books, *Hunters of the Polar North: The Eskimos* (Peoples of the Wild series). Amsterdam: Time-Life Books, 1981.

Highwater, Jamake, *Native Land: Sagas of the Indian Americas.* Boston: Little, Brown, 1986.

Hoffman, Banesh, with Helen Dukas, *Albert Einstein: Creator and Rebel.* New York: Viking Press, 1972.

Holroyd, Stuart, *Mysteries of Life.* London: Aldus Books, 1979.

Hultkrantz, Åke, *Native Religions of North America.* San Francisco: Harper & Row, 1987.

Hurkos, Peter, *Psychic: The Story of Peter Hurkos.* New York: Bobbs-Merrill, 1961.

Huxley, Francis, *The Way of the Sacred.* Garden City, N.Y.: Doubleday, 1974.

Ions, Veronica, *Indian Mythology.* New York: Peter Bedrick Books, 1986.

Jaffé, Aniela, ed., *C. G. Jung: Word and Image.* Transl. by Krishna Winston. Princeton, N.J.: Princeton University Press, 1979.

Jobes, Gertrude, *Dictionary of Mythology: Folklore and Symbols.* New York: Scarecrow Press, 1962.

Johnson, Russell, and Kerry Moran, *The Sacred Mountain of Tibet: On Pilgrimage to Kailas.* Rochester, Vt.: Park Street Press, 1989.

Jung, C. G., *Memories, Dreams, Reflections.* Ed. by Aniela Jaffé. Transl. by Richard Winston and Clara Winston. New York: Random House, 1965.

Kaufmann, William J., III, *Black Holes and Warped Spacetime.* San Francisco: W. H. Freeman, 1979.

Landes, David S., *Revolution in Time.* Cambridge, Mass.: Harvard University Press, 1983.

Lawlor, Robert, *Sacred Geometry: Philosophy and Practice.* New York: Crossroad, 1982.

Leach, Maria, ed., *Funk & Wagnalls Standard Dictionary of Folklore, Mythology and Legend.* San Francisco: Harper & Row, 1984.

Lehrman, Fredric, comp., *The Sacred Landscape.* Berkeley, Calif.: Celestial Arts, 1988.

Lethbridge, Tom, "The Master Dowser." *The Unexplained* (London), Vol. 3, Issue 31.

Levine, Robert, with Ellen Wolff, "Social Time: The Heartbeat of Culture." *Psychology Today,* March 1985.

Loewe, Michael, and Carmen Blacker, eds., *Oracles and Divination.* Boulder, Colo.: Shambhala, 1981.

Logan, Robert K., *The Alphabet Effect.* New York: William Morrow, 1986.

McClintock, Jeffrey, "Do Black Holes Exist?" *Sky & Telescope,* January 1988.

MacCulloch, John Arnott, ed., *The Mythology of All Races: Eddic.* Vol. 2. New York: Cooper Square, 1964.

McDowell, Bart, *Journey across Russia.* Washington, D.C.: National Geographic Society, 1977.

McGraw-Hill Encyclopedia of Science & Technology. Vol. 14.

New York: McGraw-Hill, 1987.

Maclagan, David, *Creation Myths: Man's Introduction to the World.* London: Thames and Hudson, 1977.

Malin, Stuart, and Carole Stott, *The Greenwich Meridian.* Maybush, Southampton, England: Ordnance Survey, 1989.

Mandel, Paul, "Seer Stalks Boston Strangler." *Life,* March 6, 1964.

Marden, Luis, and Flip Schulke, "Titicaca, Abode of the Sun." *National Geographic,* February 1971.

Marriott, Alice, and Carol K. Rachlin, *American Indian Mythology.* New York: New American Library, 1972.

Marshall, Roy K., *Sundials.* New York: Macmillan, 1963.

Meerloo, Joost A. M., *Along the Fourth Dimension.* New York: John Day, 1970.

Meisenhelder, Thomas, "An Essay on Time and the Phenomenology of Imprisonment." *Deviant Behavior,* Vol. 1, 1985.

Menninger, Karl, *Number Words and Number Symbols.* Transl. by Paul Broneer. Cambridge, Mass.: Massachusetts Institute of Technology, 1969.

Michell, John, *The Earth Spirit.* New York: Avon, 1975.

Middleton, John, ed., *Myth and Cosmos.* Garden City, N.Y.: Natural History Press, 1967.

Mookerjee, Ajit:
Ritual Art of India. London: Thames and Hudson, 1985.
Tantra Magic. New Delhi, India: Arnold-Heinemann, 1977.

Morford, Mark P. O., and Robert J. Lenardon, *Classical Mythology.* New York: Longman, 1977.

Morley, Sylvanus, *The Ancient Maya.* Stanford, Calif.: Stanford University Press, 1983.

Morris, Michael S., and Kip S. Thorne, "Wormholes in Spacetime and Their Use for Interstellar Travel." *American Journal of Physics,* May 1988.

Morris, Richard, *Time's Arrows: Scientific Attitudes toward Time.* New York: Simon & Schuster, 1986.

Mountford, Charles P., *Ayers Rock: Its People, Their Beliefs and Their Art.* London: Angus and Robertson, 1965.

Mullett, G. M., comp., *Legends of the Hopi Indians.* Tucson, Ariz.: The University of Arizona Press, 1979.

Mysteries of the Ancient Americas. Pleasantville, N.Y.: Reader's Digest Association, 1986.

Natural Wonders of the World. Pleasantville, N.Y.: Reader's Digest Association, 1980.

Navon, Robert, ed., *The Pythagorean Writings.* Transl. by Kenneth Guthrie and Thomas Taylor. Kew Gardens, N.Y.: Selene Books, 1986.

Nicholson, Irene, *Mexican and Central American Mythology.* New York: Peter Bedrick Books, 1985.

Oberg, Alcestis, "A Cave Dweller's Chronicle: Fifty-Six Days and Counting." *Omni,* June 1989.

Oldenburg, Don, "Fast Forward: Living in Artificial Time," *Health,* September 1988.

Osborne, Harold, *South American Mythology.* Feltham, Middlesex, England: Paul Hamlyn, 1968.

Parrinder, Geoffrey:
African Mythology. London: Paul Hamlyn, 1967.
World Religions. New York: Facts on File, 1985.

Pennick, Nigel, *The Ancient Science of Geomancy.* London: Thames and Hudson, 1979.

Perry, Susan, and Jim Dawson, *The Secrets Our Body Clocks Reveal.* New York: Macmillan, 1988.

Philip, J. A., *Pythagoras and Early Pythagoreanism.* Toronto: University of Toronto Press, 1966.

Phillips, Perrott:
"Against All the Odds." *The Unexplained* (London), Vol.

3, Issue 31.
"Coincidences and Connections." *The Unexplained* (London), Vol. 3, Issue 30.

Pinsent, John, *Greek Mythology.* Twickenham, Middlesex, England: Newnes Books, 1982.

Playfair, Guy Lyon, "Worlds within Worlds." *The Unexplained* (London), Vol. 3, Issue 28.

Priestley, J. B., *Man and Time.* Garden City, N.Y.: Doubleday, 1964.

Purce, Jill, *The Mystic Spiral: Journey of the Soul.* London: Thames and Hudson, 1985.

Ralph, Leslie, *Pythagoras.* London: KRIKOS, 1961.

Ralph, Martin R., "The Rhythm Maker." *The Sciences,* November/December 1989.

Rapoport, Zh. Zh., "Physical Development and Morbidity Pattern of Children in the Polar North." In *Problems of the North,* transl. by the National Research Council of Canada. Ottawa: May 1970.

Rawson, Philip:
The Art of Tantra. London: Thames and Hudson, 1973.
Tao. London: Thames and Hudson, 1987.

Rickard, Bob:
"When Fish Pour Down like Rain." *The Unexplained* (London), Vol. 1, Issue 8.
"A Wild Talent." *The Unexplained* (London), Vol. 8, Issue 87.

Rifkin, Jeremy, *Time Wars: The Primary Conflict in Human History.* New York: Simon & Schuster, 1987.

Rosenberg, Donna, *World Mythology.* Lincolnwood, Ill.: National Textbook, 1986.

Rossbach, Sarah, *Feng Shui: The Chinese Art of Placement.* New York: E. P. Dutton, 1983.

Roy, Archie:
"The Bookies' Nightmare." *The Unexplained* (London), Vol. 6, Issue 76.
"Time: The Last Frontier." *The Unexplained* (London), Vol. 8, Issue 87.
"The Waiting Future." *The Unexplained* (London), Vol. 7, Issue 84.

Rucker, Rudy, *The Fourth Dimension: A Guided Tour of the Higher Universes.* Boston: Houghton Mifflin, 1984.

Saltmarsh, H. F., *Foreknowledge.* London: G. Bell & Sons, 1938.

Schilpp, Paul Arthur, ed., *Albert Einstein: Philosopher-Scientist.* La Salle, Ill.: Open Court, 1988.

Segré, Emilio, *From X-Rays to Quarks: Modern Physicists and Their Discoveries.* San Francisco: W. H. Freeman, 1980.

Shallis, Michael, *On Time.* New York: Shockton, 1983.

Sheldrake, Rupert, *The Presence of the Past: Morphic Resonance and the Habits of Nature.* New York: Random House, 1989.

Simmons, Leo W., ed., *Sun Chief: The Autobiography of a Hopi Indian.* New Haven, Conn.: Yale University Press, 1971.

Smith, Hedrick, *The Russians.* New York: New York Times Book Company, 1976.

The Smithsonian Institution: Annual Report of the Board of Regents. Washington, D.C.: Government Printing Office, 1912.

Snyder, George Sergeant, *Maps of the Heavens.* New York: Abbeville Press, 1984.

Sorokin, Pitirim A., and Robert K. Merton, "Social Time." *The American Journal of Sociology,* March 1937.

Spielberg, Nathan, and Bryon D. Anderson, *Seven Ideas That Shook the Universe.* New York: John Wiley & Sons, 1987.

Steiger, Brad, *Mysteries of Time and Space.* Englewood Cliffs, N.J.: Prentice-Hall, 1974.

Stemman, Roy:
Mysteries of the Universe. London: Aldus Books, 1978.
"A Thoroughly Modern Medium." *The Unexplained* (London), Vol. 9, Issue 100.

Tanous, Alex, with Harvey Ardman, *Beyond Coincidence.* Garden City, N.Y.: Doubleday, 1976.

Thompson, D'Arcy Wentworth, *On Growth and Form.* Cambridge: Cambridge University Press, 1917.

"The Times of Your Life." *Time,* June 5, 1989.

Tipler, Frank J., "Rotating Cylinders and the Possibility of Global Causality Violation." *Physical Review,* April 15, 1974.

Trefil, James S., *The Moment of Creation.* New York: Charles Scribner's Sons, 1983.

Turner, A. J., *The Time Museum* (catalog). Vol. 1. Rockford, Ill.: 1981.

Valens, Evans G., *The Number of Things.* New York: E. P. Dutton, 1964.

de la Vega, Garcilaso, *The Incas.* Transl. by Maria Jolas. New York: Avon, 1961.

Ward, James A., *Railroads and the Character of America 1820-1887.* Knoxville, Tenn.: The University of Tennessee Press, 1986.

Waters, Frank, *Book of the Hopi.* New York: Ballantine Books, 1971.

Westwood, Jennifer, ed., *The Atlas of Mysterious Places.* New York: Weidenfeld & Nicolson, 1987.

Wilson, Colin:
Enigmas and Mysteries. Garden City, N.Y.: Doubleday, 1976.
Mysteries. New York: G. P. Putnam's Sons, 1978.
"Tom Lethbridge: The Master Dowser." *The Unexplained* (London), Vol. 3, Issue 31.

Winfree, Arthur T., *The Timing of Biological Clocks.* New York: Scientific American Library, 1987.

Wosien, Maria-Gabriele, *Sacred Dance: Encounter with the Gods.* New York: Avon, 1974.

Wright, Lawrence, *Clockwork Man.* New York: Horizon Press, 1969.

Zahl, Paul A., "One Strange Night on Turtle Beach." *National Geographic,* October 1973.

Zohar, Danah, *Through the Time Barrier: A Study of Precognition and Modern Physics.* London: Granada, 1983.

PICTURE CREDITS

The sources for the illustrations in this book are listed below. Credits from left to right are separated by semicolons, from top to bottom by dashes.

Cover: Art by Rob Wood, Stansville, Ronsaville and Wood, Inc. 7: Nick Nicholson/Image Bank. 8, 9: Charles Weckler/Image Bank. 10, 11: Hugh Swift. 12, 13: Kerry Keeffe/Nomad University, Bellevue, Washington. 14, 15: Bernadette Joyner/Nomad University. 17: Art by Rob Wood, Stansville, Ronsaville and Wood, Inc. 18: Mary Evans Picture Library, London. 20: Courtesy the Trustees of the British Library, London. 21: Scala, Florence. 22, 23: Rijksmuseum Van Oudheden Leyden, the Netherlands; Michael Holford, Loughton, Essex. 24: From *The Mystic Spiral, Journey of the Soul* by Jill Purce, Thames and Hudson, London, 1974, courtesy of D. M. Archer and M. G. Paish Collection, London. 25: Babji/Janata Studios, Hyderabad, India, courtesy Jagdish and Kamala Mittal Museum of Arts, Hyderabad, India. 26, 27: Pablo Bartholomew, courtesy National Museum, New Delhi. 28: Ajit Mookerjee, London. 29: Courtesy Idemitsu Museum of Art, Tokyo. 30: Österreichische Nationalbibliothek, Vienna; The National Museums and Galleries on Merseyside, Liverpool. 31: Wheelwright Museum of the American Indian, Santa Fe-Ajit Mookerjee, London; courtesy of the Royal Ontario Museum, Toronto, Canada. 33: Courtesy of the Textile Department, Royal Ontario Museum, Toronto, Canada, 914.7.8; background artwork by Time-Life Books. 34, 35: Si Chi Ko, courtesy Living Tao Foundation, Urbana, Illinois; inset Michael Holford, Loughton, Essex/The Science Museum, London. 36, 37: Robert Harding Associates, courtesy the Trustees of the British Museum, London—Museum für Völkerkunde, Abt. Amerik. Archöologie, SMPK, Berlin, photo by Dietrich Graf. 38: Österreichische Nationalbibliothek, Vienna. 39, 40: Scala, Florence. 41: Scala, Florence/courtesy Museo del Prado, Madrid. 43, 44: Larry Sherer. 46, 47: Art by Rob Wood, Stansville, Ronsaville and Wood, Inc.; artwork by Time-Life Books. 48, 49: R. T. F. Gantes/Pitch, Paris; artwork by Time-Life Books. 51: Art by Yvonne Gensurowsky, Stansbury, Ronsaville and Wood, Inc. 52: Rare Books and Manuscripts Division, The New York Public Library, Astor, Lenox and Tilden Foundations. 53: The Master and Fellows of Trinity College, Cambridge. 54: Bildarchiv Preussischer Kulturbesitz, Berlin. 55: Brown Brothers—Case Western Reserve Library, courtesy AIP Niels Bohr Library. 56: Artwork by Sam Ward. 57: Hebrew University of Jerusalem, courtesy AIP Neils Bohr Library. 59: Artwork by Lily Robins. 60: Carl Iwasaki, *LIFE* Magazine © TIME Inc. 61: National Optical Astronomy Observatories, Tucson, Arizona. 62: David Gamble/*TIME* Magazine. 65: Dr. Halton Arp, Max Planck Institut für Astrophysik, Garching/Jean Lorre, Jet Propulsion Laboratory, Pasadena; Al Freni. 67: Artwork by Gene Garbowski. 68, 69: Bildarchiv Preussischer Kulturbesitz, Berlin—AIP Niels Bohr Library/AIP Meggers Gallery of Nobel Laureates; courtesy Cern, Geneva. 71: Artwork by Stansbury, Ronsaville and Wood, Inc. 72, 73: Art by Yvonne Gensurowsky, Stansbury, Ronsaville and Wood, Inc.; artwork by Time-Life Books. 74, 75: Art by Yvonne Gensurowsky, Stansbury, Ronsaville and Wood, Inc.; artwork by Time-Life Books. 76, 77: Art by Yvonne Gensurowsky, Stansbury, Ronsaville and Wood, Inc. 79: Art by Greg Harlin, Stansbury, Ronsaville and Wood, Inc. 80: Tom Ives/*TIME* Magazine—A. Galvano/Gamma Liaison. 82, 83: Paul A. Zahl © 1973 National Geographic Society; C. G. Hampson from Annan Photo Features—Lincoln P. Brower. 84: Wally McNamee, from *A Day in the Life of the Soviet Union,* Collins Publishers, 1987. 85: Bryan and Cherry Alexander, Sturminster Newton, Dorset. 86, 87: Volker Hinz/Black Star, from *A Day in the Life of America,* Collins Publishers, 1986. 88, 89: Gerald Brimacombe/The Image Bank. 90, 91: Bruce Jackson. 92, 93: Rajesh Bedi. 94: Graham Grieves/Robert Harding Picture Library, London. 96: Pascal Marechaux, Paris; Robert Azzi/Woodfin Camp. 98, 99: Paul Chesley/Photographers Aspen, from *A Day in the Life of Japan,* Collins Publishers, 1985. 101: Archiv für Kunst und Geschichte, Berlin. 102: Akademische Druck und Verlagsanstalt, Graz, Austria. 103: Steve Krongard/The Image Bank. 104, 105: Giraudon, Paris (4)—by permission of the Folger Shakespeare Library, Washington, D.C.; Edimedia, Paris. 106: Courtesy the Trustees of the British Library, London. 107: Claus Hansmann—courtesy of The Time Museum, Rockford, Illinois (2). 108: By permission of the Folger Shakespeare Library. 109: Courtesy the Trustees of the British Museum, London; courtesy of The Time Museum, Rockford, Illinois. 110: Bettmann Newsphotos—Smithsonian Institution, Annual Report, 1911. 111: David Stoecklein/Uniphoto. 113: Art by Kim Barnes, Stansbury, Ronsaville and Wood, Inc. 114: Angus McBean, London, courtesy Parapsychology Foundation, New York—Michael Holford, Loughton, Essex, courtesy the Trustees of the British Museum. 115: Mary Evans Picture Library, London; Tad Wakamatsu, courtesy Alex Tanous—Walter Daran, *LIFE* Magazine © TIME Inc. 118, 119: From *Music Forms* by Geoffrey Hodson, Theosophical Publishing House, Adyar, Madras, India, 1976, except background artwork by Time-Life Books. 122: Bildarchiv der Österreichische Nationalbibliothek, Vienna; Mary Evans Picture Library, London. 123: Hugo Charteris. 124, 125: Mary Evans Picture Library, London. 126: Syndication International, London. 127: The Hulton-Deutsch Collection, London. 128: Gerry Cranham's Colour Library, Coulsdon, Surrey. 130: Syndication International, London/Mina Lethbridge; background Robert Estall, Colchester, Essex. 131: From *Ghost and Divining-Rod* by T. C. Lethbridge, Routledge and Kegan Paul, London, 1963. 132, 133: From *Enigmas and Mysteries* by Colin Wilson, Doubleday and Company, Inc., New York, 1976. 134: Bildarchiv Preussischer Kulturbesitz, Berlin-Pinchos Horn. 135: © Michael Holford, Loughton, Essex, courtesy the Trustees of the British Museum—Mary Evans Picture Library, London. 136: Mary Evans Picture Library, London. 137: From *Enigmas and Mysteries* by Colin Wilson, Doubleday and Company, Inc., New York, 1976.

INDEX

TIME-LIFE BOOKS

EDITOR-IN-CHIEF: Thomas H. Flaherty

Director of Editorial Resources: Elise D. Ritter-Clough
Executive Art Director: Ellen Robling
Director of Photography and Research: John Conrad Weiser
Editorial Board: Dale M. Brown, Janet Cave, Roberta
Conlan, Robert Doyle, Laura Foreman, Jim Hicks, Rita
Thievon Mullin, Henry Woodhead
Assistant Director of Editorial Resources: Norma E. Shaw

PRESIDENT: John D. Hall

Vice President and Director of Marketing: Nancy K. Jones
Editorial Director: Russell B. Adams, Jr.
Director of Production Services: Robert N. Carr
Production Manager: Prudence G. Harris
Supervisor of Quality Control: James King

Editorial Operations
Production: Celia Beattie
Library: Louise D. Forstall
Computer Composition: Deborah G. Tait (Manager),
Monika D. Thayer, Janet Barnes Syring, Lillian Daniels
Interactive Media Specialist: Patti H. Cass

Time-Life Books is a division of Time Life Incorporated

PRESIDENT AND CEO: John M. Fahey, Jr.

Library of Congress Cataloging in Publication Data
Time and space / by the editors of Time-Life Books.
  p. cm.—(Mysteries of the unknown)
  Includes bibliographical references and index.
  ISBN 0-8094-6396-2   ISBN 0-8094-6397-0 (lib. bdg.)
  1. Space and time—History.   I. Time-Life Books.
  II. Series.
  BD632.T52 1990
  115—dc20   89-20663
                                                       CIP

MYSTERIES OF THE UNKNOWN

SERIES EDITOR: Jim Hicks
Series Administrator: Myrna Traylor-Herndon
Designers: Tom Huestis, Christopher M. Register

Editorial Staff for *Time and Space*
Associate Editors: Sara Schneidman, Marion Ferguson
Briggs (pictures); Janet Cave (text)
Text Editors: Robert A. Doyle, Margery A. duMond
Researchers: Constance Contreras, Christian D. Kinney,
Stephanie Lewis, Sharon Obermiller, Carla Reissman
Staff Writer: Marfé Ferguson Delano
Assistant Designer: Susan M. Gibas
Copy Coordinators: Mary Beth Oelkers-Keegan,
Colette Stockum
Picture Coordinator: Leanne G. Miller
Editorial Assistant: Donna Fountain

Special Contributors: Lesley Coleman (London, picture research); Denise Dersin (lead research); Patti H. Cass, She-
lia M. Green, Patricia A. Paterno, Jacqueline Schaffer (research); John Clausen, Lydia Preston Hicks, Alison Kahn,
Robert A. Kiener, Harvey Loomis, Frank Leon McCoy, Susan Perry, Peter W. Pocock, Daniel Stashower, John
Tompkins, Robert White (text); John Drummond (design);
Hazel Blumberg-McKee (index).

Correspondents: Elisabeth Kraemer-Singh (Bonn), Christina
Lieberman (New York), Maria Vincenza Aloisi (Paris), Ann
Natanson (Rome). Valuable assistance was also provided
by Angelika Lemmer (Bonn); Nihal Tamraz (Cairo); Robert
Kroon (Geneva); Judy Aspinall, Christine Hinze (London);
John Dunn (Melbourne); Simmi Dhanda (New Delhi); Eliz-
abeth Brown (New York); Ann Wise (Rome); Traudl
Lessing (Vienna).

Consultants:
Marcello Truzzi, general consultant for the series, is a
professor of sociology at Eastern Michigan University. He
is also director of the Center for Scientific Anomalies Research (CSAR) and editor of its journal, the *Zetetic Scholar.*
Dr. Truzzi, who considers himself a "constructive skeptic"
with regard to claims of the paranormal, works through
the CSAR to produce dialogues between critics and propo-
nents of unusual scientific claims.

Robert L. Forward, who consulted on the time-travel essay, earned his Ph.D. in Gravitational Physics from the
University of Maryland in 1965. For thirty-one years he
worked as a research scientist at the Hughes Aircraft Research Laboratory in Malibu, California. Aside from his
current consulting work in advanced space-propulsion
technology, Dr. Forward has written extensively in the
fields of science fact and science fiction.

Robert W. Smith is a historian of astronomy who holds
joint appointments at the National Air and Space Muse-
um, Smithsonian Institution, and in the History of Science
Department of the Johns Hopkins University. Among his
recent publications are *The Expanding Universe* and *The
Space Telescope: A Study of NASA Science, Technology and
Politics.* He advised the editors on chapter 2: "The New
Visions of Science."

Other Publications:

TRUE CRIME
THE AMERICAN INDIANS
THE ART OF WOODWORKING
LOST CIVILIZATIONS
ECHOES OF GLORY
THE NEW FACE OF WAR
HOW THINGS WORK
WINGS OF WAR
CREATIVE EVERYDAY COOKING
COLLECTOR'S LIBRARY OF THE UNKNOWN
CLASSICS OF WORLD WAR II
TIME-LIFE LIBRARY OF CURIOUS AND UNUSUAL FACTS
AMERICAN COUNTRY
THE THIRD REICH
VOYAGE THROUGH THE UNIVERSE
THE TIME-LIFE GARDENER'S GUIDE
TIME FRAME
FIX IT YOURSELF
FITNESS, HEALTH & NUTRITION
SUCCESSFUL PARENTING
HEALTHY HOME COOKING
UNDERSTANDING COMPUTERS
LIBRARY OF NATIONS
THE ENCHANTED WORLD
THE KODAK LIBRARY OF CREATIVE PHOTOGRAPHY
GREAT MEALS IN MINUTES
THE CIVIL WAR
PLANET EARTH
COLLECTOR'S LIBRARY OF THE CIVIL WAR
THE EPIC OF FLIGHT
THE GOOD COOK
WORLD WAR II
HOME REPAIR AND IMPROVEMENT
THE OLD WEST

*For information on and a full description of any of the Time-
Life Books series listed above, please call 1-800-621-7026 or
write:*
Reader Information
Time-Life Customer Service
P.O. Box C-32068
Richmond, Virginia 23261-2068

This volume is one of a series that examines the history
and nature of seemingly paranormal phenomena. Other
books in the series include:

Mystic Places	*Witches and Witchcraft*
Psychic Powers	*Magical Arts*
The UFO Phenomenon	*Utopian Visions*
Psychic Voyages	*Secrets of the Alchemists*
Phantom Encounters	*Eastern Mysteries*
Visions and Prophecies	*Earth Energies*
Mysterious Creatures	*Cosmic Duality*
Mind over Matter	*Mysterious Lands and Peoples*
Cosmic Connections	*The Mind and Beyond*
Spirit Summonings	*Mystic Quests*
Ancient Wisdom	*Search for Immortality*
* and Secret Sects*	*The Mystical Year*
Hauntings	*The Psychics*
Powers of Healing	*Alien Encounters*
Search for the Soul	*The Mysterious World*
Transformations	*Master Index*
Dreams and Dreaming	* and Illustrated Symbols*